STRANGE MEDICINE

STRANGE MEDICINE

A Shocking History of
Real Medical Practices Through the Ages

Nathan Belofsky

A PERIGEE BOOK

A PERIGEE BOOK
Published by the Penguin Group
Penguin Group (USA) Inc.
375 Hudson Street, New York, New York 10014, USA

USA | Canada | UK | Ireland | Australia | New Zealand | India | South Africa | China

Penguin Books Ltd., Registered Offices: 80 Strand, London WC2R 0RL, England
For more information about the Penguin Group, visit penguin.com.

Library of Congress Cataloging-in-Publication Data

Belofsky, Nathan.
Strange medicine : a shocking history of real medical practices through the ages /
Nathan Belofsky.
pages cm
"A Perigee Book."
Includes bibliographical references and index.
ISBN 978-0-399-15995-4
1. Medicine—History. 2. Traditional medicine—History. I. Title.
R133.B33 2013
910.9—dc23   2013007162

First edition: July 2013

PRINTED IN THE UNITED STATES OF AMERICA

10 9 8 7 6 5 4 3 2 1

Text design by Kristin del Rosario

Contents

Contents

Introduction

FROM the ancient Greeks until the time of Lincoln, medicine did more harm than good, and hurt more than helped. Historian David Wootton has written, "For 2,400 years patients have believed that doctors were doing good; for 2,300 years they were wrong."

Greek doctors of two thousand years ago were at least as effective as, and probably did less harm than, the physician/astrologers of the Middle Ages, or the pompous windbags of the Renaissance, or the medical wrecking balls of medicine's "Heroic Age." Only in the twentieth century did medicine regain its stride, too late for most.

Hippocrates would have been appalled.

With due apology to historians and scholars, many of medicine's strangest ideas and dumbest procedures remain hidden from view, buried in the library stacks. This book attempts to correct this oversight and, along the way, introduces an honor roll of doctors, scientists, and medical thinkers who, however inadvertently, sent medical thought and practice hurtling backward.

This is not "history"—far from it—but the facts revealed here are true, as best we know. The book focuses on widely accepted ideas and practices by *real* doctors, not quirks or quacks, and virtually all the doctors mentioned were among the leading medical figures of their time. So, in the Middle Ages, John of Gaddesden, doctor of medicine at Oxford, really did hang cuckoos' heads from the necks of epileptic patients, and later Dr. Benjamin Rush, signer of the Declaration of Independence and treasurer of the Mint, spun his mentally ill patients like tops. Later still, Dr. Walter Freeman of Yale, the world's best-known brain surgeon, would drive ice picks into his patients' eyes with a carpenter's mallet.

Chapter One briefly surveys ancient medicine, a time of trial and error, and great ingenuity as well. Chapters Two and Three explore the medicine of the Middle Ages and the Renaissance, when leading doctors and professors dreamed up some of the most bizarre medical ideas known to humankind and then proceeded to inflict them on their patients. Chapter Four explores medicine's reckless "Heroic Age," not so long ago, when doctors went after diseases and their patients with near-comic ferocity, and predictable results.

• CHAPTER ONE •
Growing Pains

W HAT we think of as real medicine—Western medicine at least—begins with the Greeks and wise old Hippocrates, who was born on the Greek island of Kos around 460 BC. His writings and those of his followers are preserved in *The Hippocratic Corpus*, a collection of about sixty medical works.

Although most healers looked to the gods to make patients better, Hippocrates relied on the evidence of his own eyes and ears, and the touch of his hands. Above all, Hippocrates did what worked, regardless of theory or belief, and thus managed to turn the corner from magic to medicine.

Sometimes even the great Hippocrates got things wrong, big-time. Most critically, Hippocrates believed that illness was caused by an imbalance of the "four humors" (blood, black bile, yellow bile, and phlegm), a theory that would come to plague doctors and their patients until the nineteenth century.

Still, medical thought and practice flourished, with a few detours along the way. A few hundred years after Hippocrates came Galen, the great Greek anatomist living in Rome. He was forbidden by law to dissect human bodies, but his work with animals, mostly pigs, showed for the first time, from the inside, how living things actually worked.

What follows is a brief description of a few ancient medical practices and ideas, before and after Hippocrates. Some worked, some didn't, but all provide a good jumping-off point for the even stranger things to come.

The Power of Babel

[T]hey lay him in the public square, and the passersby come up to him, and if they have ever had his disease themselves . . . they give him advice, recommending him to do whatever they found good in their own case, or in the case known to them. And no one is allowed to pass the sick man in silence without asking him what his ailment is.

—GREEK HISTORIAN HERODOTUS,
WRITING OF BABYLONIAN MEDICINE

In ancient Babylonia, sick people either got better or died, with little help from their medicine or their magic.

From clay tablets fortuitously baked to stone during a fire, we know that Babylonian shamans, the *asipu*, and physicians, the *asu*, often worked together, though the *asipu* seem to have received the most attention.

Believing that some illnesses came from troublemaking gods and demons and others from a person's own bad behavior, the *asipu* determined why a person had become sick and prescribed a course of treatment. For guidance, an *asipu* might look for omens on the way to an ailing patient's house—say, the lifting of a pig's tail.

Once the *asipu* arrived, he'd root through his client's personal life. According to fragmentary accounts found in the tablets, an *asipu* might find that a patient had had sex with his mother or his neighbor's wife, or had cheated someone by saying no for yes and yes for no. Or maybe he'd had his spit stolen by a witch or been seized by the ghost of someone burned alive.

To chase illness away, an *asipu* would cast spells and chant chants. Sometimes he'd mix healing potions together in a leather bag, perhaps adding the hair of a black dog or a dirty menstrual rag. One tablet speaks of pig manure to be worn around the neck, and another speaks of a cure for teeth grinding: sleeping next to a human skull for seven days and kissing and licking the skull seven times each night.

Addud-Guppi, mother of King Nabonidus, once said:

> [O]ne hundred and four happy years . . . My eyesight was good, my hearing excellent . . . my words well chosen, food and drink agreed with me, my health was fine and my mind happy . . . [I] had my fill.

Ms. Guppi might have been a tad optimistic, but Babylonian healers did the best they could with what they had.

Die Like an Egyptian

The elite had their specialists, such as the Shepherd of the Anus and Physician of the Belly, but even for them life in ancient Egypt was difficult and short, and, as in Babylonia, there wasn't much that doctors could do about it. On the plus side, cancer wasn't much of a scourge, because few people lived long enough to get it.

From the Smith Papyrus, we know that head wounds were treated with fresh meat, and for a headache the Kahun Papyrus prescribed goose fat rubbed into the eyes, with a helping of jackass liver. A person with a toothache would have a dead mouse stuffed down their throat.

People with cataracts had hot broken glass poured into their eyes, a treatment that actually worked, and ingrown eyelashes were rubbed with bat's blood. The fat of a lion, hippo, crocodile, cat, serpent, and ibex were mixed together for baldness, along with the toes of a dog. People suffering from gout were told to stand on an electric eel.

In the Smith Papyrus, doctors performing triage gave a verdict of either "This is an ailment I can heal" or "This is an ailment I can fight with," or, for those beyond helping, "This is an ailment that cannot be healed."

Contending with a head wound, when a patient had spittle on his lips, a feeble heartbeat, and blood leaking from the ears, a doctor was to say, "An ailment with which I will contend," and apply healing ointments to the mouth. But if a patient was said to smell like the urine of a sheep and looked to be weeping, or was found paralyzed, with his phallus erect and urine dripping from it, he was to be considered beyond help. If a woman presented with a tumor in the breast leaking pus, the doctor would say, "An ailment that I will treat with the fire drill."

Sometimes all a doctor could do was prepare the egg of an ostrich, place it on a wound, and say:

> *Repelled is the enemy that is in the wound! Cast out is the evil that is in the blood . . . This temple does not fall down; there is no enemy of the vessel therein. I am under the protection of Isis . . . My rescue is the son of Osiris.*

Getting well wasn't easy in ancient Egypt, but everyone knew the afterlife would be better, and that was the best medicine of all.

Grecian Formula

> Socles, promising to set Diodorus' crooked back straight, piled three solid stones, each four feet square, on the hunchback's spine. He was crushed and died, but he became straighter than a ruler.
>
> —GREEK ANTHOLOGY XI, 120

Before Greece was even Greece, wandering sages practiced the healing arts. They thought the world was composed of earth, wind, fire, and water, and that it was their job to keep these elements in "balance" through diet, meditation, and exercise.

Hippocrates believed in the elements, the humors, and

the need for harmony in all things. But he valued the hands-on as well as the theoretical, everyday observation, and common sense. He was a craftsman, and he liked to fix things.

For whatever reason, the Greeks wrote often and at length about hemorrhoids, and Hippocrates's *On Hemorrhoids* perhaps best illustrates his no-nonsense approach to medicine, painful as it could be:

PART 1

I recommend seven or eight small pieces of iron to be prepared, a fathom in size . . . Having laid him on his back . . . burn so as to leave none of the hemorrhoids un-burnt, for you should burn them all up . . . When the cautery is applied the patient's head and hands should be held so that he may not stir, but he himself should cry out . . . [S]meared with honey and applied; the sponge is to be pushed as far up as possible.

PART 4

Having placed the man over two round stones upon his knees . . . bring it away with the finger, for there is no more difficulty in this than in skinning a sheep . . . And this should be accomplished without the patient's knowledge, while he is kept in conversation.

PART 6

[T]he hemorrhoid will separate . . . like a piece of burnt hide.

Hemorrhoids even had relevance to the lovelorn. "Love-sickness" was considered a disease, a type of "melancholy" (depression). Of melancholy, the great Galen declared, "[T]he opening of the hemorrhoids is the surest remedy."

One day Galen, an accomplished diagnostician, made a house call to Iustis, whose wife was very sick. Iustis's wife had no fever but was bedridden and acted strangely and pulled the covers over her head. During a later visit, Galen overheard a guest speak of the dreamy Pilates, a male dancer, and observed the woman's body become agitated and her face flushed. With her pulse beating wildly, the diagnosis became clear.

The Greek method for dealing with serious spinal curvature—throwing the patient off a tall building—may have been just a little too hands-on, even for Hippocrates:

> pad the ladder . . . lay the patient on it . . . tie the arms
> and hands . . . you must hoist up the ladder, either to a
> high tower or to the gable end of a house . . . let go.

Of this extreme technique, Hippocrates wrote, "[T]he ladder has never straightened anybody . . . the physicians who follow such practices . . . are all stupid."

Thousands of years before batteries, people with bad headaches were told to step on live electric eels and torpedoes (electric fish). The remedy was so well known that Plato, outwitted again by his mentor, Socrates, joked:

> [Y]ou seem to me both in your appearance and in your
> power over others to be very like the flat torpedo fish,
> who torpifies those who come near him and touch him,

as now you have torpified me . . . I do not know how to answer you.

For more serious head injuries and fractured skulls, Hippocrates employed *trepanation*. While the patient sat upright in a chair, Hippocrates drilled or chiseled through his skull, the tools becoming so hot from friction that a bucket of cold water would be kept nearby. Bone flaps and fragments would be cleaned and set, and the hole sealed with jet-black ink or pigeon's blood. Often, an otherwise dying patient would recover.

Less practical was Hippocrates's view of epilepsy. He thought it was caused by the melting of the brain and congealed phlegm in the heart. The brains of children were especially liable to rust and corrode, particularly if they stood too long in the sun or too close to the fire. Hippocrates also believed that mental illness was caused by a soggy brain, and that people with excess bile were too excitable, while phlegmy people were sullen and withdrawn. Galen once called the brain "a large clot of phlegm" but thought some of its nerves were hard to convey force of will. These were the proverbial "nerves of steel."

To Galen, women were exactly like men, only backward. A woman's private parts, he thought, faced in, when they should have been facing out:

> [T]urn outward the woman's, turn inward, so to speak and fold double the man's, and you will find them the same . . . You can see something like this in the eyes of the mole.

Appalled, Galen called it "a mutilation."

Greek doctors also thought the wombs of lonely women, stifled and starved for attention, broke free of the belly and headed upward, toward the head. Once escaped, a "wandering womb" was hard to find and even harder to catch. Plato called it "a living animal"; Aretaeus of Cappadocia "an animal within an animal." According to Aretaeus:

> [I]t moves hither and thither in the flanks . . . to the right or to the left, either to the liver or the spleen . . . in a word, it is altogether erratic.

Galen didn't think wombs wandered quite so freely, but still thought they could cause *hysterike pnix* (hysteria), which sometimes suffocated a woman. Hot, steamy sex was the best remedy, but absent that, doctors had to smoke the womb out. Believing that wombs had exquisitely sensitive noses, Hippocrates suggested "[f]umigate under her nose, burning some wool . . . sulfer and pitch. Rub her groin and the interior of her thigh with a sweet-smelling unguent."

Burned hair and squashed bedbugs were also used, but another method, not written about much but apparently commonplace, was employed as well. A smirking Martial, the Roman poet and satirist, wrote of it while describing a lonely woman afflicted with *hysterike pnix*:

> Leda told her old husband that she was hysterical and complained that being fucked is a necessity for her . . . what he no longer does, should be done. Right away the male doctors come forward and the female doctors step back, and her feet are lifted. What severe medicine!

A woman's menstruation was more predictable but still considered a force to be reckoned with, like the wind and the tides. Like *hysterike pnix*, menstruation could be deadly if left untreated. According to Hippocrates:

> [T]he girl goes crazy because of the violent inflammation, and she becomes murderous ... says dreadful things ... [visions] order her to jump up and throw herself into wells and drown.

Such was the influence of Greek medicine that later doctors believed that menstrual blood caused wine to go sour, trees to lose their fruit, and iron to rust. In an 1878 edition of the *British Medical Journal*, doctors still argued over whether menstruating women should be allowed to "rub the legs of pork with the brine pickle."

They Were Great Engineers ...

Until recently, Diaulus was a doctor; now he is an undertaker. He is still doing as an undertaker, what he used to do as a doctor.

—MARTIAL'S EPIGRAMS 1.47 (AD 100)

In ancient Rome, real men took care of themselves. If a Roman got really sick, maybe he'd hire a cobbler, a carpenter, or a blacksmith; all dabbled in medicine. Or he could go to the marketplace and hire a Greek or take one as a slave.

The first Greek to practice in Rome was the respected Arcagthus, sponsored by the government. Things didn't work out as expected, however, and Arcagthus was renamed Carnifex, meaning "executioner." It was a few hundred years before the next Greek doctor practiced in Rome.

Cato the Elder, the great statesman, hated Greeks, hated doctors, and especially hated Greek doctors. Speaking to his son Marcus, he said:

> *They are a quite worthless people... When that race gives us its literature it will corrupt all things, and even all the more if it sends us hither its physicians. They have conspired together to murder all foreigners with their physic [medicine].*

Galen, one of those Greek doctors, gave as good as he got. Referring to the Romans and assorted other hordes, he noted, "[S]ome of the barbarian tongues sound like noises that pigs, frogs and crows make... these people speak... as if they were snorting."

Roman healers weren't good at much, but they apparently did have a talent for curing warts, or at the very least making fun of people who had them. Galen wrote of a man who made a living sucking the warts off people's hands and feet, while, according to the poet Juvenal:

> *Hairy limbs and bristly arms*
> *Suggest a stern personality,*
> *But the doctor smiles as he removes*
> *The warts from your smooth anus.*

Upper-class Romans took great pride in their dental hygiene. Of a prominent Roman citizen was written:

Ignatius, because he has white teeth, is always laughing; if he be present at the felon's trial, whilst the counsel is moving all to tears, he laughs; he laughs even when everyone is mourning at the funeral pyre of a dutiful son, whilst the mother is weeping for her only child. He laughs at everything, everywhere.

For toothache, historian Pliny the Elder described rubbing one's mouth with a hippopotamus's left tooth and eating the ashes of a wolf's head. The "filth of the tail of sheep" was used to strengthen the teeth, and for toothpicks Pliny wrote of sharp bones taken from a mouse's head, and the front bones of a lizard captured in the full moon. But he cautioned against use of vulture quills, which, he said, caused bad breath. He recommended porcupine instead.

Always in love with bridges, the Romans fashioned dental plates of metal, and they filled in cavities. If all else failed, they resorted to the dreaded odontagogon, a giant tooth extractor made of lead.

A portion of the Hippocratic oath specifically bars doctors from poisoning people. This seems strange to us, but it wasn't to the Romans, who didn't like doctors but did like poison and doctors who knew how to use it. The poet Juvenal, an acute observer of the Roman upper crust, wrote that anyone who hoped to get anywhere had to be good at poisoning. As proof, he cited the faux pas of one woman who, upon poisoning her husband, found that he had swallowed the antidote beforehand; she had to stab him instead. Of course, husbands also poisoned their wives, and mothers poisoned their ungrateful children.

Nero, acclaimed for having fiddled while Rome burned, was so happy with his poisoner, Locusta, that he supplied her with pupils for her own poisoning school. Business was so good that *praegustatores* (tasters) formed their own union. The Roman emperor Claudius was poisoned by his physician, and doctors were routinely hired by persons of high standing to kill other persons of high standing. Some doctors skipped the middleman and killed people themselves.

The Romans were big on wool, said to have "awesome powers," but cabbage held their highest esteem. Cato recommended eating it, or better yet, drinking the urine of someone who had just eaten it.

A meatier alternative was hepatoscopy, the study of sheep livers, by *haruspex*, the field's high priests. Examining organ folds was serious business to the Romans, who founded a school to teach it. At a certain point, certain *haruspex* became so exalted that they constituted a threat to the state. Apollonius of Tyana, hoping to divine the best way to overthrow the emperor, was said to have sacrificed a boy for his liver. Roman leaders banned the practice, on pain of death. Still, it was *haruspex* Spurinna who said to Caesar, "Beware the Ides of March," as good a diagnosis as any.

Medicine's Dark Ages

THINK medieval, and think serfs, crusades, and bringing out the dead. But in elite universities, pampered,
purple-robed physicians spent their days lost in thought
and contemplation, debating obscure points of philosophy
in Latin.

Medieval physicians developed a peculiar reverence
for the books and medicine of the past. In particular, they
worshipped Hippocrates and Galen, whose writings, it was
believed, contained all the medicine any doctor needed to
know, for all of time. Medical thinking froze in its tracks.

Above all, medieval physicians developed a studious
disdain for the practical. It was their job to talk to patients,
not to treat them. The actual touching of a person's body
was to be avoided, like the plague.

Medieval physicians did talk a good game. Writing in
the twelfth century, John of Salisbury said, "[W]hen I hear
them harangue, I am charmed, and almost persuade myself that they can raise the dead. There is only one thing

that makes me hesitate; their theories run directly counter to one another."

Into the breach stepped hands-on surgeons, such as Henri de Mondeville, and barber surgeons, both looked on by the physicians with disdain. Craftsmen like these set bones, pulled teeth, and did surgery. By default, they became the real doctors of the age.

In the Stars

In 1348, with his kingdom decimated by the Black Death, King Philip VI of France turned to the medical professors of the University of Paris, who, after extensive study, issued a formal report. The Black Death, they determined, had begun on March 20, 1345, at one p.m. Three higher planets in the sign of Aquarius, they discovered, had aligned and corrupted the atmosphere:

> The conjunction of Saturn and Jupiter brings about the death of peoples ... The conjunction of Mars and Jupiter causes great pestilence in the air ... Jupiter, a warm and humid planet, drew up evil vapors from the earth and water, and Mars, being excessively hot and dry, set fire to these vapors.

Surely, in the face of such preordained catastrophe, medical experts such as those at the University of Paris couldn't be expected to halt the spread of the disease.

By the late Middle Ages, astrology, in both theory and

practice, was on the syllabus of Europe's leading medical schools. Statutes were passed requiring that doctors carry the latest charts and horoscopes in their medical bags.

In contrast to a doctor's own hit-or-miss observations, astrology could predict with near-perfect accuracy the exact time when a patient should be treated or a procedure carried out. Migraine, for example, was best treated on April 3, but blindness on April 11. In 1437, at the University of Paris, bitter controversy broke out over precisely when to prescribe a laxative. To avoid future scandal, doctors were warned to account for "every day, every hour and fraction of an hour."

Even the great de Mondeville made sure the time was right for his surgical procedures. Before even attempting surgery to the head, the exact phase of the moon had to be considered:

[T]he humors are agitated at that time and they wax as the light waxes on the body of the moon. The brain waxes in the skull as the water raises in the river . . . thus the membranes of the skull rise up and consequently come nearer to the skull, so that they would be more easily damaged by the surgical instruments.

Medical astrology protected doctors as well. In 1424, a London man sued three surgeons for botching his thumb surgery. After due consideration, the case was dismissed by a panel of judges. The initial injury, it was discovered, had occurred on January 31, when the "moon was consumed with a bloody sign, to wit, Aquarius, under a very malevolent constellation."

The Chirurgeons

In medieval times people suffered for their sins, and many learned to accept nothing less. Being hurt or diseased was a great way to suffer, and so was having surgery.

One of the best chirurgeons (surgeons), Henri de Mondeville, warned that any surgeon who refused to inflict as much pain as possible on his patients, even unnecessary pain, would become a laughingstock:

> It is dangerous for a surgeon . . . to spare his patient anything . . . [even] on the occasions when one can dress the wound gently . . . all ordinary uneducated people . . . distrust and mock surgeons who work gently, saying they are timid, weak and inexperienced . . .

De Mondeville continued:

> So it is! Those surgeons who treat sufferers and dress their wounds roughly and mercilessly . . . and have no more pity on them then they would on dogs . . . are held now a days to be noble, expert and resolute fellows.

Bringing on the pain wasn't a stretch in the Middle Ages—the best a patient could hope for was a sharp knife, a plank of wood to bite on, and a few strong men to hold him down. With no anesthesia, procedures were so frightful that friends and bystanders often fainted, and surgeons' assistants sometimes ran away. Of a patient's weak-kneed friends, de Mondeville wrote, "[S]ometimes a higher fee may be got from persons present fainting and breaking their heads against wood and the like than from the principal patient."

To keep a patient in place, one medieval textbook suggested he be "turned up-so down upon a dishe, or upon the knees of some strong servant . . . bounden strongly with the nekke." Another recommended:

> [H]ave a sturdy assistant sit on a bench with his feet on a stool. The patient sits on his lap with [his] legs bound to his own neck . . . or [have him] lie down on a bench or plank . . . let him be tied . . . with three bandages . . . restraining his hands and arms.

Big procedures were deadly, and even accomplished surgeons often just stood aside, to let God's will be served. When gambling with risky surgery, the smart chirurgeon made sure he got paid first and afterward moved on to the next village, so as not to overstay his welcome.

A few brave souls sliced away with gusto. When nobleman Giovanni de Pavia was wounded in 1276, a respected physician saw his intestines hanging out. A bit short on bedside manner, the doctor exclaimed, "He's a dead man." The famous William of Saliceto was summoned. William washed the patient's intestines in wine, stuffed them back into his body, and stitched them in place. De Pavia survived, married, and had children.

In the tenth century, Ali ibn Abbas al-Majusi, Persia's great physician, perfectly captured the painful realities of medieval surgery. In his instructions for a simple tonsillectomy, he wrote:

Have the patient sit opposite you facing the sun and order him to open his mouth. Order one servant to hold the head of the patient from behind, and another servant to press down the tongue . . . pull out the tonsil with a hook . . . without pulling out with it any membranes or structures . . . cut the tonsil out from its roots with scissors . . . stop the bleeding.

Cuckoo's Nest

Tickling a person out of depression made perfect sense to readers of John of Gaddesden's *Rosa Medicinae*, written in 1314. A doctor of medicine at Oxford, Gaddesden modestly proclaimed, "As the rose overtops all flowers, so this book overtops all." Readers were inclined to agree, and they made the book a best seller for centuries to come.

For the mentally ill, Gaddesden suggested:

Tie their extremities lightly and rub their palms and soles hard; and let their feet be put in salt water . . . and pull their hair and nose and squeeze the toes and fingers tightly, and cause pigs to squeal in their ears . . . open the vein of the head, or nose, or forehead, and draw blood from the nose with the bristles of a boar.

Put a feather, or a straw, in his nose, to compel him to sneeze . . . and let human hair or other evil smelling thing be burnt under his nose . . . let a feather be put down his throat ... and shave the back of his head.

In other chapters, Gaddesden recommended bleeding before a long journey, boiling a dead dog for paralysis, and that parents of poisoned kids mix goose droppings into their food without telling them.

He also cautioned that yawning could be dangerous: "I once saw one of my household who ... yawned so often, and opened his mouth so widely in doing so, that he dislocated his jaw." For toothache, Gaddesden applied a needle bathed in the goo of "a many footed worm which rolls up in a ball when you touch it."

Gaddesden warned patients, especially aspiring lovers, to wear clean underwear. For epilepsy, he roasted a cuckoo and blew its powder up a person's nose. If that didn't work, he hung the cuckoo's beak around his patient's neck.

Ring of Fire

Being sick was the easy part for victims of the hot cautery. Essentially a branding iron, the dreaded instrument was used to stanch hemorrhage, clear out dead tissue, and as an all-around pick-me-up. It could cure just about anything, from hernia and epilepsy to excess phlegm and bad skin.

William of Saliceto claimed the cautery should be used

year-round, not just in the spring. He thought iron was best because it retained its shape while white hot, but that more pliable gold irons were good for small, delicate areas, such as the eyes, and decorative as well. Roger Frugard liked to use them for depression: "[W]e perforate the cranium to allow the escape of the offensive matter . . . while the patient is restrained in bonds," he wrote.

A firm hand was required. Abu al-Qasim Khalaf ibn al-Abbas Al-Zahrawi (Albucasis), the great Islamic surgeon, described a typical procedure:

> [H]ave his head shaved; then seat him cross-legged before you, with his hands on his breasts . . . Then heat up an olivary cautery. Then bring it down upon the marked place with one downward stroke . . . If you see that some bone is exposed . . . then take your hand away; otherwise repeat . . . till the amount of bone I have mentioned is exposed.

For a case of extreme swelling in the twelfth century, a doctor recommended:

> [C]arefully burn two spots on the forehead with the round cautery, and two on the back of the head with the broad cautery, and two spots on the temples, and one under the lip, and one in the hollow of the throat, and two under the collarbone, and three spots . . . on the chest . . . and two more on one hip and two on the other, and another two under the anklebones.

Albucasis knew the iron could be fatal and suggested using it only on those strong enough to survive the shock.

Bruno of Longoburgo found the iron helpful for headaches but applied it with caution:

> *When you use the cautery on the head let not the instrument remain on the cranium too long lest the brain be cooked and its membranes shriveled.*

Moon Mad

Bald's Leechbook is a medical textbook, most likely from the ninth century. Only one copy exists, in London's British Library. The book speaks of a cure for "elfsickness," a mysterious disease thought caused by invisible tiny elves shooting tiny arrows. Victims would waste away, or freeze like statues.

> *For elfsickness . . . go on Wednesday evening when the sun is setting where you know dwarf elder to be growing . . . go away, go back to it when day and night divide . . . no matter what frightful thing or man should come towards you say no word to him before you go to the plant which you marked the previous evening . . . dig out the plant, let the knife remain, go back as quickly as you can . . . wash it and make it into a drink . . . let him drink the drink afterwards, it will soon be better for him.*

For headache, a person was to tie a stalk of crosswort (an herb) to the head with a red kerchief, and "for that one be moon-mad, take a dolphin's hide, make it into a scourge, beat the person, he will soon be better."

"Bloodletting is to be avoided for a fortnight before Lammas [the festival of the wheat harvest]," the *Leechbook* advised, "and for thirty-five days afterwards, because then all poisonous things fly and injure men greatly."

For a "wrenched skull" (apparently a stiff neck), "lay the man out flat, drive two pegs at the shoulders, then lay a board crosswise over the feet, then strike thrice with a hammer, it will go away shortly."

The Real Doctors

Elite medieval physicians talked the talk, but it was surgeons that actually treated patients, sometimes with striking ingenuity.

For broken ribs, one surgeon fed his patient gaseous foods, such as beans. This would expand and cushion the stomach. According to the *Rogerina*, written by surgeon Roger Frugard around 1180, tiny skull fractures were diagnosed by having a patient crack a nut open with his teeth or by having the patient shut his mouth, plug his nostrils, and blow hard and seeing what came out. For stomach wounds, Frugard wrote, "[I]f the intestines are cold, cut in half a live animal and put it in the intestines themselves and leave it there until they are warmed up and softened."

Not all medieval surgeons practiced missions of mercy. Of castration, one wrote:

> [W]hen the Potentates of the world need men to guard
> their wives [eunuchs] . . . there are two methods: one is
> by crushing and the other is by cutting. In the first pro-
> cedure have the patient sit over hot water until the tes-
> ticles are soft and [fall] down. Then squash them with
> your hands until they are soft and cannot be felt.
>
> [Y]ou use the knife in one of two ways: in one you
> remove the testes and the shaft of the penis, in the other
> you remove only the testis.

But this was the exception, not the rule. To heal a
wound, Albucasis made an ointment of shredded frog,
minced crabmeat, and lizard; to close it, John of Mirfield
used spiderwebs as stitching. Albucasis also put hungry
African ants on the edge of a wound, leaving their heads
behind to bind it. All three techniques worked, and the
African ant trick is still used today.

The medieval crossbow was considered barbarous, even
for the Middle Ages. In 1139, Pope Innocent II barred its
use against Christians, though apparently everyone else
was fair game. The crossbow shot pointy metallic bolts
that hit hard, penetrated deep, and caused wounds that
could overwhelm even the best chirurgeon.

Having a strong man whack the bolt out with a heavy
iron hammer was one surgical strategy. If that didn't work,
ingenious medieval surgeons employed another, more extreme procedure, using a second crossbow to undo the
damage of the first.

The technique, apparently common, is illustrated in the
Elche panels, a series of drawings from the thirteenth century. In the third panel, a crossbow is attached to a pillar

standing a few feet from the victim, who has a bolt stuck in his neck. The cord of the crossbow would be attached to the bolt, pulled taut, and released as if to fire, hopefully pulling the bolt out along with it.

In the panels, the bolt remained in the neck and the patient was doomed. But then, attended by the Virgin and two angels, he seems to have recovered. Here on earth, the surgeon de Mondeville reported great success using the technique.

The Gold Standard

Through this science I can show you the reasons of the whole universe.

—MEDIEVAL MEDICAL WRITER,
ABOUT THE EXAMINATION OF URINE

In ancient times, urine was a prophylactic, a health drink. John XXI, the only medical doctor ever to become pope, drank it religiously until the ceiling he designed himself fell on his head, killing him. Galen wasn't a big fan of urine therapy—he couldn't stand the smell—but did suggest drinking "gold glue," the urine of an innocent boy stirred in a copper pot.

By medieval times, it was believed that only urine, nicknamed "the liquid window," could truly unlock the body's secrets. In good sun, a medieval physician would examine a sample three times and determine whether the urine was yellow or green or red or purple, or, if over-cooked by the stomach, black. Taking into account wind, weather, and planetary alignment, he'd then determine his patient's health. The best results were obtained from a urine flask shaped like a human bladder, decorated with precious jewels.

Most critically, examining the urine spared a doctor the burden of having to touch a patient's body, or even look at it.

As time passed, the analysis of urine assumed even more importance. In the twelfth century, French physician Gilles de Corbeil wrote the poem *On Urine*. It was 347 lines long, and medical students had to memorize it. In urine, de Corbeil claimed, he and his protégés could find things like grease, pus, scales, and sand.

The switching of urine samples became a big problem for reasons unknown, but few could slip one past Arnald of Villanova, physician, astrologer, and alchemist:

> [T]o the individual who brings the urine . . . keep your eyes straight on him or his face . . . if he wishes to deceive you . . . the color of his face will change, and then you must curse him forever and in all eternity . . .
>
> If you have a competitor whom you believe to be a shameless crook, be careful . . . perhaps he will stir up the urine for you and you will not be able to form a certain judgment from it.

Substituting cheap wine for urine was another trick. But the clever Arnald, who invented brandy, was always one step ahead:

[A]ct as if you were going to blow your nose, whereby you put the finger that has been dipped in [the sample] on or next to your nose; then you will smell the odor of wine ... [If wine], get away and be ashamed of yourself.

Bedside Manner

In the sixth century, a dying Austragild, wife of Guntram, king of the Burgundians, made her husband swear an oath to execute her two physicians, Nicholas and Donatus. Upon her death, he did.

To Henri de Mondeville, a patient's state of mind was critical to their health, and, consequently, his as well.

On a house call, de Mondeville told doctors to bring along a violin player or someone to tell jokes. To avoid a big argument, the wife was kept out of the room. "[E]very woman," he wrote, "seems to think that her husband is not as good as those of other women."

Another medieval manuscript suggested entering the room with a "hilarious countenance," shooting the breeze about dogs, horses, and falcons, and saying, "Hey there, what do you say? What sort of fun are you having?"

According to de Mondeville, a doctor should also trick his patient into thinking things were better than they seemed. "False letters may be written relating the decease of his enemies, or those from whose death he expects advantage," he graciously explained. As an example, if a patient aspired to leadership in the church, "he should be told that the bishop is dead and he is elected."

When dealing with family members, the physician had to be on his best behavior. He had to exude an aura of competence, earned or not; flatter the family, no matter how detestable; and pretend to know, and care, about the patient himself. The twelfth-century *De Adventu Medici* (*The Doctor's Visit*), attributed to Archimatthaeus, advised:

> On the way learn as much as possible from the messenger, so that if you discover nothing from the patient's pulse or water, you may still astonish him . . . praise the beauty of the country and the house . . . or the liberality of the family . . . Do not be in a hurry to give an opinion, for the friends will be more grateful for your judgment if they have to wait for it.

If invited for a sit-down dinner, a visiting physician was not to gorge himself, no matter how good the food, and had to inquire after the sick patient, at least occasionally. Looking desirously at a man's wife, daughter, or handmaid was of course forbidden.

The Doctor's Visit was later corrupted into a handbook for less legitimate practitioners, falsely attributed to the good and decent Arnald of Villanova. "Suppose you know nothing, [then] say there is an obstruction of the liver," it advised. If, as was likely, the patient said the pain was elsewhere, the doctor was to insist "liver obstruction" in a loud voice. "[P]atients do not understand it, which is very important," the book said.

Celebrated anatomist Gabriel Zerbi, in his *Caute-lae Medicorum* (*Hints for Doctors*), published sometime in the fifteenth century, instructed doctors not to hang dirty bandages from their house, or dance too much, or be the beaters of musical instruments. They were also to steer clear of murderers and—perhaps asking too much of a medieval doctor—were prohibited from being pompous.

Endless Love

By the seventh century, eminent physicians were arguing over the best cure for lovesickness. All agreed, however, that keeping the brain sufficiently moist was absolutely critical. To achieve the desired humidity, doctors would force a lovesick man to smell the menstrual cloth of his beloved or inhale the stinking embers of her burned feces.

Two hundred years later, Rhazes of Persia enumerated the life cycle of lovesickness with great precision. First, a person's eyes hollowed out, then the tongue developed pustules, then the body shriveled. The afflicted would babble and break out in blisters. Toward the end, a doomed patient would howl like a wolf and die.

In the tenth century, the great Persian physician Abu Ali al-Husayn ibn Abd Allah ibn Sina (Avicenna) finally located the biological seat of lovesickness—the middle ventricle of the brain. As was true for Galen before him, his greatest diagnostic triumph involved simple detective work:

[He] placed his hand on the patient's pulse, and men-
tioned the names of the different districts [until] the pa-
tient's pulse gave a strange flutter. Then [he] repeated the
names of different streets of that district . . . till he
reached the name of a house at the mention of which
the patient's pulse gave the same flutter . . . Thereupon
he said: This man is in love with such-and-such a girl, in
such-and-such a house, in such-and-such a street, in
such-and-such a quarter: the girl's face is the patient's
cure.

If all else failed, Avicenna suggested hiring a woman, preferably an old one, to trash-talk the patient's beloved.

A Regular Guy

While medieval doctors waxed philosophical, workaday surgeons such as Guy de Chauliac set bones and made house calls. As more "learned" physicians fled from the Black Death, Guy de Chauliac, like many other surgeons, stayed behind to treat patients the best he could.

De Chauliac's multivolume textbook, the leading tome of its time, was hit or miss, trial and error. He accurately described mastectomy as "a fearsome enterprise," to be used only as a last resort, but also instructed surgeons not to operate on a fractured skull when the moon was full. "[T]he brain is puffed up against the dura matter," he wrote. He wisely recommended feeding chicken soup to the sick, but, in *Serpents as Remedies*, suggested using

Tyrian viper whenever possible. For bad breath, he advised a brisk rubdown with rabbit's blood.

For those over forty, he recommended bleeding three times a year, not to exceed six pounds. A special supplement suggested that bleeding be done only when the weather was good and the moon was in ascension. Priests in the choir were to have their bleeding only in the spring, to maintain their voice.

De Chauliac's concern with diet may have been ahead of its time, if a bit crude:

> *A person can become so obese by overeating fatty foods that he can barely walk, and cannot reach to clean his anus. His belly is so protuberant that he cannot remove his own shoes. His breathing is impaired and he is called Fatty.*

De Chauliac also wrote about impotence, which, sorcery and castration aside, was usually rooted in the psychological. If physical causes could be ruled out, he would hire an "experienced" woman to shower a troubled couple with wine and massage and to talk to them suggestively for three days.

The Tooth-Man

Traveling "tooth-pullers," wearing pointy hats and trophy-teeth necklaces, competed with jugglers and acrobats on Europe's carnival circuit. Drums and music would mask, or perhaps accompany, the screams of their patients. More established tooth-men advertised with a string or two of rotten teeth hung outside their shop.

A good tooth-man secured teeth with gold wire, filled cavities, and fitted dentures, perhaps made of cow bone. But for the most part he was cautious, wary of the risk of infection.

If simply yanking out a bad tooth with fingers or a length of string didn't do the job, a tooth-man would use his pelican, a tool resembling a pelican's beak. Firmly placing its claw over the tooth and twisting from the side, he'd extract the tooth, hopefully leaving the surrounding gum and bone intact.

Bloodletting was recommended for toothache. So was the burning of nerves with acid or a hot iron and rinsing with molten gold. For "tooth worms," a tooth-man would place lighted candles in a patient's mouth or smoke them out with smoldering roses. Burning the skin behind the ears was also recommended.

Understandably, medieval people did what they could to avoid a trip to the tooth-man. "Chewing sticks," to clean the teeth, were popular, and so were toothpicks. By the late Middle Ages, toothbrushes, shared by the entire family, were probably used by the wealthy, though their boar bristles would break off and fall down the throat. The twelfth-century *Trotula*, a series of books written by a female physician, suggested that women rinse their teeth with good wine, to avoid "stinking of the mouth."

On Injuries in Those
Who Suffer Torture
with Whips and by Being Hung
from Their Limbs

Treating tortured bodies was all in a day's work, judging from the collected works of renowned thirteenth-century surgeon William of Saliceto.

In his *Surgery*, Chapter 25 is matter-of-factly titled "On Injuries in Those Who Suffer Torture with Whips and by Being Hung from Their Limbs."

It begins, "When someone suffers beatings with rods, switches, thongs, etc., or when he is strung up or racked by his limbs . . ." It continues, "[Or is] tortured by basti-nado until he becomes unconscious and his limbs are engorged . . . and he is dragged down by the weight of his blood . . ."

William then described how to treat these apparently routine injuries. He suggested, for example, wrapping a torture victim in the still-warm skin of a recently flayed horse. Afterward, he recommended relaxing baths and a low-fat diet.

Roger of Parma discussed treating victims struck by the clapper of a bell or a really big key, and for arrow wounds he suggested inserting bacon along the wound's path.

In either the thirteenth or fourteenth century, cele-brated surgeon Jehan Yperman wrote *The Surgery of Mas-ter Jehan Yperman*. One chapter, "Head Wounds Inflicted by the Sharp Corners of the Blades of Cleavers, Knives or

Hatchets, or by the Points of Swords," suggested that the surgeon first look to see if the wound had penetrated the brain's outer membrane, and, if it had, to give up. Another chapter, "Stab-Wounds to the Chest," covered injuries caused by darts, clubs, "[k]nives, daggers, swords, picks, pitch-forks [and] javelins."

"Wounds of the Face . . . Caused by Projectiles" described how arrows shot upward from the base of a castle wall were more deadly than those shot from the top down, while "When an Ear Is Partially Sliced Off" cautioned surgeons not to block the ear canal. "That would make you a laughingstock," he wrote.

Though clearly expert at treating penetrating wounds, Yperman also dispensed more general surgical advice. Concerning harelips, he insisted they were caused by "the imagination of the mother during sexual intercourse," not, as was commonly believed, a mother eating rabbits during pregnancy.

Guy de Chauliac was more discreet about torture. He did write of "pulling at the body as by a rope or a chain" but called being stretched on the rack "extension," and he spoke of neck injuries caused by "falling" from the end of a rope.

Like doctors before him, de Chauliac recommended wrapping torture victims in the warm fleece of a newly killed animal, but he also suggested burying broken patients in a mound of fresh horse manure. His parting advice was, "If he is dead . . . do not attempt to treat."

Know the Ways

Hildegard of Bingen was the twelfth century's leading life-style guru, and a religious visionary. With a mind for the ages, she was also a brilliant naturalist. She described the life cycles of thirty-seven types of fish swimming by her house and identified at least seventy-two kinds of birds, though one, with the body of a lion, was the mythical griffin.

Hildegard preached moderation in all things, exercise, and a good diet. Wholesome spelt was a favorite, along with boiled hedgehog. Butter was to be avoided, along with peacock and falcon. Hildegard hated junk food—it "spread slime in the stomach . . . like a rotting manure pile"—but didn't begrudge the occasional treat, such as cookies made with gold.

Hildegard's view of anatomy was a little more dubious. In books such as *Scivias* (*Know the Ways*), she claimed that the four bodily humors began in the brain and went to the lungs, spleen, and liver, which was connected to the ear. In men, humors went downward to "the virile organ" and ended up in the legs. Women had tubelike passages in their skulls for menstruation.

Hildegard thought excess burping and hiccups were precursors of cancer. Depression was triggered by a waning moon, which caused a person's liver to be perforated by tiny holes, "like cheese." Bad weather caused postnasal drip, "so that a harmful slime collects." Running too fast could shrink the testicles and send toxic phlegm to the brain.

Hildegard had her remedies. For jaundice, she knocked a bat senseless, then tied it to the loins. Tea made from the dried liver of a lion eased indigestion, and on the genitals of oversexed men she rubbed an ointment of sparrowhawk and sap from a tree. For epilepsy, she baked a cake made of mole's blood, duck's beak, and the feet of a female goose. Water in which a skinned mouse had been boiled made a nutritious broth, and if you collected enough pelts, she noted, there'd be enough material for a nice coat. As a preventive measure, Hildegard suggested a good bleed in the convent bloodletting house.

In matters religious, Hildegard remained humble, teaching that humankind was just a small part of a much bigger cosmos. As she wrote, only the pelican, looking into a person's body and soul, truly knew what God had planned for them.

Mondeville on Money

Let doctors call in clothing fine arrayed
With sparkling jewels on their hands displayed . . .
For when well dressed and looking over-nice,
You may presume to charge a higher price;
Since patients always pay those doctors best
Who make their calls in finest clothing dressed.

—From the Regimen Sanitatis *of*
Salerno *(twelfth or thirteenth century)*

Treating a Turkish pasha, the physician Solon took a large fee and allowed his patient to rest. The pasha died unexpectedly, and Solon wisely fled the jurisdiction. But the pasha's friends caught up to him on the Dalmatian coast and sawed Solon, and his son, in half. When King John of Bohemia's eye doctor didn't cure his cataract, the king had him sewn into a sack and thrown into the river.

Henri de Mondeville, along with virtually every other prominent medieval surgeon, warned colleagues to avoid the "hard cases," and hard clients too. "[H]e should never mix himself up with [the] desperate ones," de Mondeville wrote. Of patients unwilling to pay, Gilles de Corbeil, the twelfth-century urologist, wrote:

> [P]romises are carried by the wind . . . While pain torments the patient, his love of giving is most fervid; it is then that you must draw the contract . . . When the disease improves cold avarice returns to the fore, the will to give cools down, the physician becomes a burden and his work is importyune and unpleasant for the patient, who will try to discredit his role and force him to go away empty-handed.

The Doctor's Visit, the twelfth-century textbook, even counseled doctors to be careful when taking the pulse, which would quicken when the patient thought about his fee.

Of all the surgeons, de Mondeville wrote the most about business matters. "[The] principal preoccupation of the surgeon is to be paid," he believed, and "I have never found anyone so rich, or even honest . . . who was ready to pay what he had promised, unless obliged and convicted." To doctors who assumed wealthy clients would pay for drinks and a lavish dinner, de Mondeville warned, "[I]t will be cheaper to get your dinner at an inn, for such feasts are usually deducted from the surgeon's fee."

A bit extreme was advice from a doctor of Italy's prestigious School of Salerno. If a patient didn't pay the bill, "Contrive that he shall take alum instead of salt with his meat; this will not fail to make him come out all over [with] spots."

Despite his tough talk, the prosperous de Mondeville had a good heart and, like many surgeons of his time, treated the poor for free.

The Trotula

The *Good Housekeeping* of its time, the Trotula was a set of three texts written in the eleventh or twelfth century. At least one volume was written by a woman named Trotula, a physician practicing out of the School of Salerno. The books address a wide array of "women's issues," from teeth whitening to holiday recipes, but mostly concern sex and reproductive matters.

According to the Trotula, a wandering womb—"the wild beast of the forest"—could suffocate a woman, especially a sexless woman. Like Hippocrates, she suggested stuffing burned wool or stinky leather up the nose and rubbing the vagina with sweet-smelling ointments.

For vulva tightening, "so that even a woman who has been seduced may appear a virgin," Trotula suggested a rub made of snake blood and pomegranate bark, or dwarf olives and plantains boiled in rainwater. Boiling an old shoe and breathing in the fumes was good for excessive menstruation. In a section on women who "can't endure the male organ on account of its length or largeness," Ms. Trotula counseled against the use of ground glass.

Concerning health and diet, if a woman was too fat to have children, Trotula told her to slather on a cream made of cow dung and a "very good wine," then sit in a hot sauna until she turned green. An equally fat husband or boyfriend was to be swathed in the same mixture and then buried in the sand next to the ocean.

A Day in the Life

Torture, plague, and starvation made the headlines, but medieval citizens also dealt with more ordinary health issues, such as head lice and floaters in the eye.

If someone was choking on a fish bone, a doctor would tie a sponge to a string, lower it down the throat, and reel in his catch. Bee-stung patients were rolled in the sand, and walnut shells or long needles with hooks were used to dig peas or worms out of the ear.

To treat vermin in the scalp, thirteenth-century surgeon Jehan Yperman would baste a person in mercury and make him stand near a fire. The vermin would die from toxic fumes, hopefully before the patient did.

Of treatment for "melancholy," Arnald of Villanova is said to have written the following:

My master ordered a pig to be hanged to the bed's head of a Neapolitan soldier who was lethargic; the perpetual clamor of the beast terrified him [so] that he could not sleep.

In another case, Arnald supposedly treated an apathetic patient by shaving his head and smearing it with honey, forcing him to swat away the flies.

For floaters, many patients apparently insisted on the hot cautery iron, but surgeons also fished them out with a metal hook.

Gargling with goat urine was one remedy for a person who accidentally swallowed a leech. If that didn't work, a patient would abstain from water to make sure the leech was thirsty. Then, a hollow bronze tube would be shoved down the patient's throat. Next, a red-hot cautery iron would be shoved down the tube. Finally, a nice, cold glass of water would be placed in front of the patient, and the thirsty leech would be grabbed when it made its move.

In his *Thesaurus Pauperum* (*Treasure for the Poor*), Pope John XXI compiled everyday treatments from the most prominent physicians. For gout, a doctor was to skin a very fat puppy; stuff him with equal parts vulture, goose, fox, and bear fat; boil him; and feed him to the patient. Alternatively:

[I]f you like, take a frog when neither sun nor moon is shining; cut off its hind legs and wrap them in deer skin; apply the right to the right and the left to the left foot of the gouty person.

Bad Doctor

Usamah ibn Munqidh was a Muslim warrior who fought against the crusaders. In roughly 1175, he met a doctor, Thabit, who told him of his brush with Frankish (German) medicine. Both found it appalling, even by medieval standards. Munqidh wrote:

Thabit was absent but ten days when be returned ... He said:

They brought before me a knight in whose leg an abscess had grown; and a woman afflicted with imbecility. To the knight I applied a small poultice ... and the woman I put on diet and made her humor wet. Then a Frankish physician came to them ... He then said to the knight, "Which wouldst thou prefer, living with one leg or dying with two?" The latter replied, "Living with one leg."

The physician said, "Bring me a strong knight and a sharp ax." A knight came with the ax ... Then the physician laid the leg of the patient on a block of wood and bade the knight strike his leg with the ax and chop it off at one blow. Accordingly he struck it ... but the leg was not severed. He dealt another blow, upon which the marrow of the leg flowed out and the patient died on the spot.

He then examined the woman and said, "This is a woman in whose head there is a devil . . . Shave off her hair." . . . Her imbecility took a turn for the worse. The physician then said, "The devil has penetrated through her head." He therefore took a razor, made a deep cruciform incision on it, peeled off the skin . . . until the bone of the skull was exposed and rubbed it with salt. The woman also expired instantly.

Thereupon I asked them whether my services were needed any longer . . .*

---

* Apparently, the Frank legal system wasn't much better. Munquidh reported, "They installed a huge cask and filled it with water . . . They then bound the arms of the man charged . . . tied a rope around his shoulders and dropped him into the cask, their idea being that in case he was innocent, he would sink . . . This man did his best to sink . . . but he could not do it. So . . . they pierced his eyeballs with red-hot awls."

· CHAPTER THREE ·

The "Renaissance"

Down to the river though I know the river is dry.
—Bruce Springsteen, "The River"

WITH a few glorious exceptions, the great achievements of the Renaissance passed medicine by. Vesalius reinvented anatomy, Ambroise Paré revolutionized surgery, and William Harvey showed how blood circulated through the body. But their heroic efforts, not to mention vast leaps in scientific knowledge, had little effect on everyday doctors and their practices. Mainstream medicine, still following the old ways, remained ignorant and pompous as ever.

As before, hands-on surgeons did what they could, under the most trying of circumstances. From Hieronymus Fabricius (1537–1619) we have this account:

I was about to cut the thigh of a man forty years of age, and ready to use the saw and cauteries. For the sick man

no sooner began to roare out, but all ranne away, except
only my eldest Sonne, who was then but little, and to
whom I committed the holding of the thigh . . . and but
that my wife then great with child, came running out . . .
and kept hold of the Patient's chest.

Meanwhile, eminent medical professors argued over whether to draw blood from the same or opposite side of a person's bad humors. And when doctors dissected dead people's bodies and discovered that they looked different from Galen's anatomical drawings, derived from pigs, they blamed the bodies and sided with Galen.

Patients knew the score. The playwright Molière wrote, "Doctors know how to speak Latin, know all the ancient Greek names for the diseases . . . But for curing them, they know nothing at all." At Renaissance carnivals, send-ups like Dr. Braggart would spout, "He who is sick cannot claim to be well," and, "A walking man isn't dead." Even Leonardo Fioravanti, a well-known doctor of the sixteenth century, noted the clash between Renaissance thinking and Renaissance reality: "When I saw anatomy done, I never saw phlegm, choler (yellow bile) melancholy or vital spirits, or any of the other fabulous things that physicians dream up."

Trifling details like this didn't stop the enlightened doctors of the Renaissance from repeating the mistakes of the past—or adding to them.

The Royal Treatment

When King Charles II died unexpectedly in 1685, his loyal subjects demanded details. The king's medical staff, anxious to avert blame, released a number of diaries and journals. These would prove, once and for all, that the king had received the best care possible.

On February 2, the king woke up feeling sick. His shave was interrupted, and a pint of blood was drawn. Elite doctors were summoned by messenger, and another eight ounces of blood were taken, using suction cups.

His Royal Majesty was forced to swallow antimony, a toxic metal. He vomited and was given a series of enemas. His hair was shaved off, and he had blistering agents applied to the scalp, to drive any bad humors downward.

Plasters of chemical irritants, including pigeon droppings, were applied to the soles of the royal feet, to attract the falling humors. Another ten ounces of blood was drawn.

The king was given white sugar candy, to cheer him up, then prodded with a red-hot poker. He was then given forty drops of ooze from "the skull of a man that was never buried," who, it was promised, had died a most violent death. Finally, crushed stones from the intestines of a goat from East India were forced down the royal throat.

King Charles died on February 6, 1685.

The Beak Doctors

During the plague years, many if not most "doctors" hid from their patients or simply fled. Those brave souls who stayed behind, mostly surgeons, tried to protect themselves from contagion with birdlike masks, featuring beaks up to a foot and a half long. Some called them beak doctors.

The beaks functioned like crude gas masks and were filled with aromatic herbs and spices to overcome "bad air." Eyeholes in the mask were covered with red-tinted glass, to stare down the evil eye.

Of the beak doctors, one seventeenth-century poem read:

> *When to their patients they are called*
> *In places by the plague appalled . . .*
> *Their hats and cloaks of fashion new*
> *Are made of oilcloth, dark of hue,*
> *Their caps with glasses are designed*
> *Their bills with antidotes all lined,*
> *That foulsome air may do no harm*
> *Nor cause the doctor man alarm.*

The ensemble could include an ankle-length overcoat smeared with wax or fat; as accessories, gloves, boots, and a tight hat; and amulets of dried blood or ground-up toads. A wooden cane, called a tickle stick, would allow the doctor to poke and prod at a distance.

The Weapon Salve
and the Sympathetic Powder

The weapon salve's key ingredient was moss growing from the skull of a thief recently hung from chains. But something else distinguished the weapon salve from all others. The soothing ointment was applied not to the wound but to the weapon that caused the wound. Elite seventeenth-century doctors believed in it and used it on patients.

Doctors came to realize that asking a mortal enemy to turn over his lucky sword following a fight to the death took some nerve, so, after thinking things over, they decided that a wooden replica of the original weapon would do just as well.

Meanwhile, the Catholic Church was coming down hard on witches and witchcraft. Even clergyman William Perkins, affiliated with the "moderate-puritan" wing of the Church of England, said, "The most horrible and detestable monster . . . is the good witch."

Not wanting to be on the wrong side of that argument, doctors scrambled to show the "scientific" basis of the weapon salve. A leading European physician, Daniel Beckher, declared that the ointment's power came from a doomed man's coagulated animal spirits. Shortly after coalescing with his vital essences, Beckher discovered, these spirits would travel upward. While the man was being hanged, they'd burst out the circumference of his cranium. Beckher also relied on the work of the renowned Goclenius the Younger, who, though accused of consorting with demons, had the last word, as usual, with his *actio in distan* (action at a distance) theory.

Johann Hartman, with his "private chemical college," also pitched in.

Even better than the weapon salve, and more widely employed, was the sympathetic powder, which was synthesized, crystallized, and then titrated. The powder's leading proponent was Sir Kenelm Digby, who in 1658 lectured about it before "the Learned Men of Montpellier," site of the world's leading medical school. Digby described the case of Mr. J. Howell, who was injured when he tried to break up a duel. Howell was cured when, in another room, Digby dipped his underwear into the sympathetic powder. Another time, a carpenter was accidently cut with an ax. When the tool was sprinkled with powder, the man regained his strength. Doctors were puzzled when he took a turn for the worse. Sure enough, the offending ax had fallen from the nail on which it hung, shaking off the powder.

With witch hunting still in vogue, Highmore of Oxford announced that the powder worked through a confluence of atoms, fluids, and pores, catalyzed by zaphyrian salt. For the technically minded, he noted that "sharper angles grate the orifices of the capillary veins [causing] an efflux of blood."

So powerful was the sympathetic powder that a 1687 pamphlet proposed it be used to solve the daunting longitude problem, the nemesis of sailors for centuries. A person onshore would be given the bandage of a wounded dog, and the dog sent to sea. When the landlubber dipped his bandage into the powder at a prearranged time, the dog aboard the ship would yelp, meaning, "The Sun Is Upon the Meridian in London." By comparing London time with his own, even the most clueless captain would know exactly where he was.

The salve and the powder really did work, in their own way. They kept a doctor's hands off the patient and let the wound heal by itself.

The Great Paracelsus

With his embrace of chemistry and rejection of the old ways, Paracelsus shook the very foundations of Renaissance medicine. He also believed in elves, nymphs, water people, sylphs, will-o'-the-wisps, fire people, and gnomes that dwelt in mines where precious healing metals were found.

Paracelsus sneered at medical books and the people who read them, so in his own 7,500 pages of writings he wrote things like "the more learning the more perverted" and "not even a dog killer can learn his trade from books." Most of all, Paracelsus enjoyed writing about how he hated doctors, and for this the doctors hated him back.

In 1527, Paracelsus posted a public notice:

The famous Doctor Paracelsus, City Physician, will speak at High Noon tomorrow . . . He will . . . touch upon the Ignorance, the Avarice and the Strutting Vanity of the Doctors of Basel.

In case anyone missed the point, he also threw the doctors' beloved medical classics into a bonfire and accused them of being frauds and money-grubbers. The good doctors of Basel and even Paracelsus's own medical students chased him out of town.

Thus began years of wandering, and Paracelsus honed his craft. He learned that the stars caused disease through their exhalations and that orchids cured venereal disease because they looked like testicles.

A wizened Paracelsus rejected traditional anatomy and surgery and instead offered "the elementals"—half spirit, half human, and capable of curing illness and disease. He coined the word *gnome* to describe those foot-high, wrinkled old men with long white beards who walked through rock. Gnomes dressed in brown, emerged from tree stumps, and were good for bone-setting. The most powerful of the elementals, the salamanders, were often seen peering into houses. They cured liver disease.

Paracelsus far outshone his peers, according to Paracelsus:

> I am *Theophrastus [Paracelsus] ... I can prove to you what you cannot prove ... every little hair on my neck knows more than you and all your scribes, and my shoe buckles are more learned than your Galen and Avicenna.*

Paracelsus died in 1541, loved by some and hated by most.

The Living Dead

Of healing the living with the flesh of the dead, Paracelsus wrote, "[I]f doctors were aware of the power of this substance, no body would be left on the gibbet [gallows] for more than three days." A favorite recipe of his was:

> Take moss growing on a skull exposed to the weather, and of human fat each two ounces, of mummy and human blood each half an ounce ... Make an ointment and preserve it in a box.

In the fifteenth century, physician Marsilio Ficino, noting that women called "screech-owls" sucked the blood of babies to good effect, was moved to ask:

> Why shouldn't our old people ... likewise suck the blood of a youth? ... They will suck, therefore, like leeches ... they will do this when hungry and thirsty and when the moon is waxing.

In *The Marrow of Physick* (1669), Thomas Brugis wrote:

> [Take] a Mans Skull that hath been dead but one yeare, bury it in the Ashes behind the fire, and let it burne until it [is] very white ... then take off the uppermost part of the Head ... beat it ... grate a Nutmeg, and put to it ... the blood of a Dog ... mingle them altogether and give the sick to drinke.

Not to be outdone, a medical-minded Franciscan monk published a recipe for a healing marmalade, made of human blood. "[S]tir it to a batter with a knife ... pound it ... through a sieve of finest silk."

Paracelsus's protégé, noted chemist Johann Schroeder, had his own recipe:

> *Take the fresh, unspotted cadaver of a redheaded man . . . aged about 24 who has been executed and died a violent death . . . cut the flesh into pieces and sprinkle it with myrrh and just a little aloe. Then soak it in spirits . . . let the pieces dry in a shady spot.*

Renaissance pharmacists kept human ingredients in stock and, for impulse buyers, made them highly visible and affordably priced, like magazines near the cash register. German officials recommended that pharmacists shelve no fewer than twenty-three different varieties of human body parts. A model inventory from 1652 included dried flesh, *mumie* ("the menstruation of the dead"), "human grains," *menschenfleisch* (marinated human flesh), human fat, *usnea* (moss growing on a human skull), and spirit of bone, all to be applied topically or mixed in a very stiff drink.

Naturally, human blood was at a premium, the fresher the better. All through the eighteenth century, Dutch and German doctors encouraged epileptic patients to push their way to the scaffold where felons were being beheaded, to catch a few precious drops of blood from their still-quivering bodies.

First Impressions

Be it a leaping cat or a favorite painting, what a pregnant woman saw could be stamped onto her newborn, according to the *doctrine of maternal impressions*. Usually the exchange from mother to fetus wasn't for the better.

In the thirteenth century, a Roman noblewoman gave birth to a boy with fur and claws. Authorities found the defect was caused by her excessive admiration of a bear painting hanging in her living room. Pope Martin IV ordered that all statues and paintings of bears be destroyed.

In 1550, Swedish archbishop Olaus Magnus, one of the most respected scientific minds of his day, declared, "[T]here is one misfortune that many women meet with in pregnancy, either by eating or by leaping over the head of a hare; they bear children with a hare mouth."

Ambroise Paré, the father of modern surgery, believed in maternal impressions, especially after learning of a woman who had held a toad in her hand and borne a child with the head of a frog. In another well-documented case, a pregnant woman was known for her great love of herring—she ate more than 1,400 of them. The woman gave birth to a boy, not a sardine, but witnesses said the first words out of his mouth were, "I want some herring."

Thomas Batholin of Denmark, discoverer of the lymphatic system, was the doctrine's foremost proponent. He knew of a pregnant woman startled by a cat jumping out from under her bed. She gave birth to a girl with a feline head. Danish king Frederik IV planned to build a hospital for the handicapped; he didn't want to cure the lame, just keep them out of sight of expectant mothers. In the eighteenth century, medical advisors to the court of King George found that a woman's love of rabbit meat caused her to bear rabbits, and the medical literature noted a seafood lover who had a child shaped like a mussel.

Doctors eventually became skeptical of the doctrine, leading to a dramatic showdown in a German courtroom. In 1790, a woman was accused of adultery, having married a black man and given birth to a white child. The woman testified that a painting of a pale-skinned man hung on the wall of her apartment and that she stopped often to admire its artistry. She was convicted anyway.

Cutting Edge

Soon to be the world's greatest anatomist, a young Vesalius, seeking objects to study, embarked on a distinguished career in grave robbing. One day he and a friend came across the find of a lifetime. A notorious robber had been chained to the top of a high stake and left to bake in the sun:

In that way there had been a dish cooked for the fowls of heaven, which was regarded by them as a special dainty. The sweet flesh of the delicately roasted thief... had been elaborately picked and there was left suspended on the stake a skeleton dissected out and cleaned by many beaks with rare precision.

Awash in bones and bodies, Vesalius practiced and prospered, and his anatomical drawings turned the medical world upside down. But Vesalius made powerful enemies when he exposed the mistakes of the old-timers, such as Galen's belief that blood was generated by the liver. A few missteps, he likely feared, and it would be his bones being picked clean by the vultures.

Vesalius's chief nemesis was his old teacher Sylvius, said to be so mean that he smiled only on the day his cat died.

Sylvius wrote, "It would have been easier to cleanse the Augean stables than to remove . . . [this] monster . . . suppress him so that he may not poison the rest of Europe with his pestilential breath."

Vesalius was ultimately vindicated, but his heresies continued to haunt him. He was falsely accused of dissecting a man whose heart was still beating, and he eventually grew so fed up that he gave up anatomy altogether. Even as a well-heeled advisor to King Charles V, the former grave robber was mocked by his medical colleagues. They did everything they could to stay away from actual bodies, dead or alive, but were more skilled than Vesalius in matters of philosophy and rhetoric.

Vesalius is said to have died aboard a ship. He supposedly left strict instructions to preserve his body and bury it on land, to avoid his being torn apart by the creatures of the sea.

Treasure Hunt

According to the *doctrine of signatures*, God's visitation of misery and illness upon the world was mercifully tempered by his planting of secret clues that, discovered in time, would allow people to recover from them. Of this precious gift, one grateful theologian wrote that God had given humanity "the means of curing the disease which he provided."

The doctrine's central premise was that plants and herbs that looked like body parts or diseases could be used to cure them, if a sick person or his doctor were clever enough to connect the dots. Thus walnuts, which "Have the perfect Signatures of the Head," were created by God to cure headaches, and liverwort to treat the liver.

God's little game was played by some of the world's leading doctors, particularly Paracelsus, and was later popularized by the *Signatura Reum* (*The Signature of All Things*). That the book was written by a shoemaker didn't seem to matter, and the theory was found in respected medical textbooks until the nineteenth century.

Moonlighting

They didn't give up their day jobs, but executioners in Germany and Austria also moonlighted as doctors, their forensic experience perfectly suited to healing the wounded and curing the sick.

In contrast to more theoretical-minded physicians, executioners were used to working with people's bodies—living, dead, and in transition. They knew all about pain and had a good grasp of things like anatomy, circulation, and bodily functions. Often they would nurse a torture victim back to health so he could be tortured some more.

Patients came to appreciate the executioner's personal touch, and they became popular among rich and poor alike. An ability to kill one day and cure the next reflected well on the profession, but not every executioner could become a healer. Like doctors, executioners had to be licensed. In *Defiled Trades and Social Outcasts*, historian Kathy Stuart writes of one applicant who claimed he was a good horse doctor, and others who admitted to being too old to chop off people's heads. Executioners also took an oath swearing they never murdered anyone and not to work with Jews. Frederick the Great insisted that his executioner/healers pass a written exam.

Still, there was a backlash, especially because, in a competitive marketplace, executioners had first dibs on much-desired body parts. *Armsunderfett* (poor sinner's fat), for example, was sold to drugstores by the pound, and possibly peddled from street stalls as well.

Executioner medicine declined by the seventeenth century, but the core practice remained vibrant for centuries to come.

Lovesickness, the Book

In 1610, Jacques Ferrand wrote his famous *Treatise on Lovesickness*. The book, though happily not the doctor, was burned at the hands of the Inquisition.

According to the treatise, love was insidious:

> [I]f some serpent comes breathing into the ears a few tempting words, proposes some dalliance or other with her coaxing and wheedlings, or some basilisk [serpent] comes along casting lascivious looks, winking, and making sheep's eyes, those hearts very quickly allow themselves to be seized and poisoned . . . In just this way the demon of love and lechery plays it at the beginning with those he means to take.

Ferrand thought lovesickness arose when heat from a fevered lover bleached blood into sperm. Allowed to fester, this fluid would "[r]emain in its reservoirs, turn corrupt, and from there-by means of the backbone and other secret channels—send a thousand noxious vapors to the brain."

To treat the problem, Ferrand supported the usual clamping of a metal ring to the foreskin but opposed lining love-struck patients with thin plates of lead. And "[B]ernard of Gordon goes too far, I think, when he says the lover should be spanked and whipped until he begins to smell bad all over."

Ferrand wasn't entirely dismissive of generations past:

Avicenna, Prince of the Arab physicians, in his chapter on love, Bernard of Gordon, Arnald of Villanova, and several other modern physicians teach that in order to prevent erotic melancholy in someone who starts meddling with love, we should make him fall in love with some new friend, and that when he starts making soft eyes at her, work at making him hate her too, and to fall for a third, and so on several times over until he is tired of love altogether.

In his *Surgical Cures for Erotic Melancholy*, Ferrand took a more hands-on approach. He advocated bloodletting three or four times a year, especially if a lover was "sufficiently corpulent, well-fed and plump."

Sometimes more drastic action was necessary:

But if the patient's imagination is already deranged, I prefer to have a medial vein opened . . . if the blood flows

> black, thick and heavy a good deal can be drawn . . . I
> continue by bleeding the ankle, particularly in women
> patients, some of whom suffer specifically from hysteri-
> cal suffocation or uterine fury . . .
>
> Or even more to the purpose, I would provoke the
> flowing of the hemorrhoidal veins . . . "the opening of the
> hemorrhoids is the surest remedy both for the prevention
> and the curing of any melancholy disease.

Ferrand also suggested shaving down an overly long clitoris and burning a woman's thighs with acid. If a case of lovesickness was so urgent that it threatened to turn into lycanthropy (werewolfism), he recommended bleeding the veins of the arms until complete heart failure ensued, then cauterizing the front of the head with a searing hot iron.

The Visible Sin

Sometimes evidence of a person's sinful nature was as plain as the nose on their face, or what remained of it.

Use of the nose as a scarlet letter dates back at least to the time of the Egyptians, when randy judges had their noses (and ears) cut off during the Great Harem Conspiracy. Adulterers in ancient Rome and Greece had their noses cut off, and duels and wars also took their toll. To deter sexual assault, women sometimes cut off their own noses. In the ninth century, a convent of French nuns amputated their noses to escape sexual assault at the hands of invading Saracens. They were massacred, but not raped.

During the Renaissance, many people lost their noses to syphilis. Nose reconstruction, with ancient but effective techniques imported from India, became big business.

In 1442, Italian poet Elisio Calenzio wrote a friend:

> *Orpianus, if you wish to have your nose restored, come here . . . Branca of Sicily, a man of wonderful talent, has found out how to give a person a new nose, which he either builds from the arm or borrows from a slave . . . you shall return home with as much nose as you please. Fly!*

But face fixing eventually ran afoul of religious authorities, who wanted the noses of syphilitic sinners left exactly as they were. Sometimes things got done, in secret.

In 1460, Heinrich von Pfolspeundt wrote a letter titled "To Make a New Nose for One Who Lacks It Entirely, and the Dogs Have Devoured It":

> *[If] a person comes to you, whose nose has been devoured, and you wish to make a new nose for him, let no one watch, and make him promise you solemnly to conceal how you wish to heal him . . . find a trustworthy person who will also promise to conceal the matter . . . [to] help the patient eat and drink . . . And the room wherein he lies must also be locked.*

In the sixteenth century, Gaspare Tagliacozzi developed a procedure allowing arm-to-face skin grafts to "take." He had patients wear a strange-looking helmet that pinned their arm to their nose, forcing them into a weeks-long salute. The church believed that Tagliacozzi, a mere mortal, was tampering with God's will, especially God's will to maim and disfigure. Since many of the noses Tagliacozzi fixed had collapsed because of "the visible sin of syphilis."

Tagliacozzi became an outcast. He was attacked by medical giants like Paré and, in death, satirized by Voltaire. He was excommunicated by the church. His brilliant surgery mostly disappeared, and stayed disappeared, for hundreds of years.

Anatomy Day

Even a dead body could put on a show, and attending public dissections in the sixteenth and seventeenth centuries was considered both fun and fashionable. Anatomy day, for those lucky enough to score a ticket, was a festive time, with speeches, processions, and music. An Italian playbill from 1597 notes a procession of lute players and the audience's "tumult and stomping."

To accommodate growing demand, dissection "theaters" went up all over Europe. In Padua, Italy, the Las Vegas of anatomy, anatomist Hieronymus Fabricius became such a draw that a new, roomier forum had to be built. The venues themselves were elaborate, decked out in fine woods and expensive artwork. Ushers kept order and gate crashers, peeking through windows and from behind columns, were ejected by bouncers. Anyone with a ticket was welcome—visiting dignitaries, medical students, salted fish dealers, even Jews.

Questions from the audience were permitted, but laughing too hard was discouraged. Audience members passed around body parts for inspection, though the taking of souvenirs was prohibited, and some jurisdictions enforced stiff fines. Proceeds from the box office went to fund lavish banquets, or to pay the hangman.

It Has Been Proved

The Book of Medicines, intended to combat the quackery so prevalent in sixteenth-century Germany, was written by Oswaldt Gabelthouer, court physician to the Duke of Württemberg. Gabelthouer was so confident that he issued a kind of guarantee—following each "cure" is the Latin word *probatum*, which means "it has been proved." Among his remedies:

Epilepsy: Skin a small mouse; remove its entrails except the lungs and liver; burn . . . mash to powder . . . a tablespoonful every morning; half portion for children.

[For] very serious cases . . . Take the right eye of a wolf, the left of a shewolf; dry, and hang about the neck of the patient, who must wear them for three months continuously, during which time he must [not] bathe.

Nausea: [the author] has tried it with the greatest success for forty nine years of practice . . . Take the brains of a fox . . . bake it, and give it to the patient on an empty stomach.

For a woman . . . take a young female dog, cut it

open, *carefully secure its gall bladder, prick it and let the gall ooze . . . and give to the patient to drink . . . hold your hand over her mouth . . . she is likely to vomit.*

Dizziness: . . . a few drops of the blood of turtle doves in a small glass of wine . . . hellebore and coltsfoot . . . refrain from garlic, onions, sauerkraut.

Insomnia: Take the grease from an ass's ear—it matters not what kind of ass—and apply to the patient's temples.

Insanity: Cut the patient's hair close to his head . . . then cut a ram's liver in two and bind it while yet warm on the patient's head.

Complete Insanity: . . . take a freshly baked loaf of bread; remove the soft inner part and replace with a complete ox-brain; bind on the patient's head and it will cure his brain and restore his mind.

Remembrance of Things Past

An outbreak of angst swept the world in 1688, carrying with it a teary-eyed longing for the way things used to be.

The wave of sentiment was first discovered by Swiss physician Johannes Hofer. He successfully treated a young man who, while studying in Berne, missed his native Basel. Another of Hofer's patients, a visiting country girl, had a terrible fall and lapsed into a coma. Upon waking in a strange bed, she reportedly asked for her parents and said, "I want to go home, I want to go home."

Similar incidents led to the classification of a new disease, *nostalgias* (nostalgia), from the Greek *nosos*, meaning "native land," and *algos*, meaning "suffering and grief." By 1700, the illness was accepted by doctors all over the world, its causes and treatment much discussed in medical books and journals.

In his seminal treatise, Hofer claimed the disease was caused by vital forces moving along oval tubes connecting the center of the brain to the body, made worse by sticky sludge obstructing the organs. Symptoms included constipation, heart problems, fever, and not laughing at jokes that were funny, Dr. Hofer wrote, and in more serious cases, resulted in death.

Besides a change of scenery, aggressive bloodletting was recommended, with or without leeches. In 1789, however, French doctor Jourdan LeConte claimed nostalgia was best treated with pain and terror. He cited the success of a Russian general who, facing an outbreak, buried two of his troops alive.

Later, some would claim that nostalgia was just another type of "pathetic insanity," but continued research ensured the disease's place in medical textbooks. Most well known was the case of a homesick British soldier, cured upon learning that he was about to be released from military service.

During America's Civil War, surgeons wrote often about the disease, finding that "country bumpkins" suffered more than city slickers, already used to the "perambulating life." Army doctors reported 5,213 cases of "nostalgia" in 1861, with about sixty deaths.

In 1899, England's Royal College of Physicians declassified nostalgia as a disease.

The Great Pox

Syphilis, called "the French disease," or, to the Russians, "the Polish disease," ravaged Europe in the fifteenth century. To the leading medical thinkers of the time, the epidemic was preordained by planetary alignment, and just punishment for the wicked.

The best-known pox doctor was Girolamo Fracastoro. He honored the affliction with an epic poem, twenty years in the writing and thirty-six pages long. The poem described the tortuous journey of a little shepherd boy named Syphilus ("piglover" in Latin), whose disgusting behavior insulted the god Apollo, and who reaped the consequences.

Real life wasn't much better. In the sixteenth century, German poet and playboy Ulrich von Hutten contracted syphilis:

> *Boils that stood out like Acorns, from whence issued such filthy stinking Matter . . . The Colour of these was of a dark Green and the very Aspect as shocking as the pain itself, which yet was as if the Sick had laid upon a fire.*

Italian physician Gaspar Torella applied freshly flayed pigeons. Other doctors used hot irons, knives, and drills. But mercury salve and heat was the treatment of choice.

Fracastoro basted his patients like turkeys. One doctor wrote, "The stench of frying fat was through the air." The goal, physically and metaphorically, was to purge the body, so that "all the rottenness leaves the body with the drops of sweat." Sometimes patients would salivate eight pints of black phlegm a day.

Against the advice of friends, von Hutten refused to commit suicide. He wrote bitterly of abandonment at the hands of his aristocratic physicians, who, as usual, fled town. He also wrote of treatment by the clueless but brave surgeons who stayed behind:

> [C]austics to burn out the ulcers . . . mercury . . . shut up in a steam room . . . twenty and thirty days . . . ulcerated gums . . . teeth would loosen and drop out. Many sufferers preferred death.

Patients often got their wish, "in the grip of a most appalling agony." Von Hutten undertook the cure eleven times and died at age thirty-five.

So desperate was the need for treatment, and so shaming the attitude against sufferers, that the French government granted a nobleman named Le Febure exclusive rights to sell chocolate laced with mercury:

> A husband can take his chocolate in the presence of his wife without her suspecting a thing; indeed she herself can take it without realizing that she is swallowing an anti-venereal remedy; by this innocent means, then, peace and concord can be maintained in the household.

As it turned out, the mercury itself was eating away at syphilitic patients. By the seventeenth century, German medical candidates took an oath swearing they wouldn't become "poison-mixers and murderers" by prescribing it.

Mercury remained widely used, however, even in the United States. Dr. Nathaniel Chapman, the first president of the American Medical Association, wrote:

> *If you could see what I see ... emaciated to a skeleton, with both plates of the skull almost completely perforated ... the nose half gone, with rotten jaws and ulcerated throats ... You would exclaim, as I have often done ... horrid, murderous quackery.*

Bosum Serpents

Medieval doctors took the story of Nero very seriously. Curious about how babies were born, he supposedly had his mother killed and cut in half. Unsatisfied, he then had his doctors create a potion allowing him to bear a child himself. Nero drank the potion, and a frog grew inside him, and he vomited the frog, soaked in blood and hideous to behold. Nero then instructed his doctors to keep the monstrous spawn in a vaulted room and to feed and care for it.

A thousand years later, in 1561, savvy Ambroise Paré was a little more skeptical. When a Parisian woman caused an uproar by claiming a snake had crawled into her stomach, he threatened to give her a strong laxative. The woman recanted, though a few days later she was again seen standing by the city gates, inviting men to pat her belly and feel the snake inside.

Other Renaissance doctors had no trouble believing

that toads and lizards, called *bosum serpents*, lived in their patients' bellies.

Frog fever crested in the seventeenth century in Altenburg, Germany, when a woman supposedly vomited dozens of frogs, toads, and lizards. Her painful spasms were corroborated by leading medical experts, who came from all over Europe just to see her. Johannes Schenkius, in his *Observations Medicorum Rariorum*, confidently declared that frog spawn adhered to the lining of the stomach and that colonies bred in the gastrointestinal canal.

In 1732, Carl Linnaeus, a medical doctor and the father of modern taxonomy, treated a Finnish woman with three frogs in her stomach. He said he heard them croaking. Linnaeus prescribed liquid tar, but the woman drank aquavit instead, and the hungover frogs happily stayed put. In his *Iter Lapponicum*, Linnaeus expressed great concern over the problem, warning that the frogs and toads teeming in Finland's brooks and streams could "tear and torture [a] poor patient."

In the nineteenth century, a doctor in the Russian royal court, Martin Wilhelm Mandt, treated a man who swallowed a snake. Mandt felt the snake wriggle in his patient's stomach and, using a stethoscope, heard it gurgling. Mandt prescribed a strong laxative. Two days later, the patient handed Mandt a chamber pot that contained a large snake. With irrefutable evidence, Mandt published his astounding findings in medical journals all over Europe.*

---

* Around the same time, Nathaniel Hawthorne wrote "Egotism, or the Bosum Serpent," a short story about a man who, harboring shameful secrets, thought a serpent was growing in his belly. Literary experts still view the story as allegory, but Hawthorne apparently knew better.

A Cabinet of Curiosities

Doctors loved to display medical curiosities to their patients—the more grotesque the better. Few, however, had the desire or the means to create a genuine "cabinet of curiosities," a collection of medical and scientific wonders from around the world.

In the sixteenth century, Ulisse Aldrovandi, the "Bolognese Aristotle," created one of the world's first great cabinets. His was said to feature a dragon, maybe two, and came with a companion book, *Monstrorium Historia*. The cabinet belonging to malaria and plague expert Athanasius Kircher displayed a vomiting eagle, a perpetual motion machine, and a perpetual screw, along with the usual medical freaks and curiosities.

Wormian bones, which fill the gaps in a baby's head, are named for Ole Worm, aka Olaus Wormius. In the seventeenth century, however, he was best known for his Wormianum in Copenhagen. Visitors to the Worm called it a *Wunderkammern*, or "room of wonders." The museum's most impressive acquisition was probably the Vegetable Lamb of Tartary—half animal, half plant, and bearing sheep as its fruit. Also on display were skulls from a race of prehistoric giants and a coffee bean from the New World. Within its walls Dr. Worm conducted research as well; there he discovered that the bird of paradise had legs and that the unicorn did not exist.

Curating his own cabinet, Frederik Rusch, an eighteenth-century pioneer of the lymphatic system, specialized in giant dioramas. They featured dead people pickled in a secret preservation solution, most likely pig's blood.

Rusch's daughter helped with the design, decorating the dioramas with seashells, flowers, and fine lace.

Rusch's dioramas were designed to teach a moral lesson. A shriveled corpse, for example, might cling tightly to a string of pearls, signaling the futility of earthly riches. A lover of the philharmonic, Rusch also conducted a "corpse orchestra." The pickled bodies, instruments in hand, would be carefully arranged into a symphony or perhaps a string quartet. Rusch would back his dead bodies with live musicians and serve light refreshments.

More modern "cabinets" skipped the moral and went right to macabre. Medical museums in nineteenth-century New York City, open to "gentlemen" only, featured spear-throwing savages from the "Island of Senegambia," "instructional" images of beheadings, and "the most revolting specimens of cutaneous disorders." An 1840 article in the *Provincial Medical and Surgical Journal*, titled "An Untoward Result of Drunkenness," breathlessly reported:

> *The Museum of the Royal College of Surgeons, in Dublin contains the picture of a man whose face was eaten away by a pig, while he was left lying, in a state of intoxication. The entire nose, both cheeks, and parts of his face, in fact all the most eatable parts, of his face, were chewed off by the animal; nevertheless, the wounds all healed; but of course . . . [with] extensive destruction of the soft parts. His principal regret lay in the unavoidable use of his tobacco-pipe.*

Opened in 1858, the Mutter Museum of the College of Physicians of Philadelphia, "America's finest museum of medical history," still dazzles medical voyeurs with a two-headed baby, a nine-foot colon from the Human Balloon, assorted freaks, and an ovary the size of a soccer ball.

Smile

Ambroise Paré, the great surgeon, liked to dabble in dentistry but was careful to distinguish himself from the tooth-pullers of yore. Paré wrote of Picard, who promised to extract one bad tooth but accidentally yanked three good ones instead. Threatened with the filing of a complaint, Picard told the patient he was lucky to have gotten three teeth removed for the price of one.

Meanwhile, Europe's top medical minds were churning out dental textbooks and treatises. One doctor wrote of the connection between teeth and eyesight, another of cankers caused by the south wind. In 1595, Dr. Jacob Horst, professor of medicine at the University of Helmstadt, wrote of a boy born with a gold tooth. Dr. Horst attributed this phenomenon to the boy having been born when the sun was in conjunction with Saturn, in the sign of Aries, and believed the special tooth portended a time of happiness and plenty for all. But because the tooth was ominously located on the lower left side of the jaw, the professor was concerned that the coming golden age was to be preceded by a dark era, marked by tumult and upheaval.

The boy's story remained mired in controversy for a hundred years until Duncan Lidell, a Scottish professor teaching in Germany, wrote his seminal *Tractatus de dente aureo pueri Silesiani* (1695) and accompanying letters to his medical colleagues.

Lidell first noted that Horst had assigned the wrong astrological sign to the boy, and that proper astrological analysis all but ruled out the possibility of a gold tooth. More significantly, Lidell wrote of a certain conversation he had with Herr Baron Fabianus, the *rector magnifucus* of the college where Lidell taught. The esteemed baron revealed to Lidell that, years before, a nobleman had insisted that the golden boy bare his teeth. When the boy refused, the nobleman, now drunk, stabbed him in the mouth with a dagger. The doctor who treated the injury found in the boy's mouth a worn-out, and man-made, gold inlay.

In the seventeenth century, Johann Strobelberger wrote his well-received *Complete Treatise on Gout in the Teeth*, sixty-seven chapters long, and Lazare Rivière, of the prestigious Montpelier Medical School, discovered that teeth were connected not to the eyes but to the ears. This finding was confirmed by Renaldus Fredericus, who noted that if one chewed on the tip of a pole stuck in the ground in the middle of the night, the footsteps of an approaching person could be heard.

Carved into an ancient Babylonian tablet are the words:

> *[T]he Worm came crying before Shamash . . .*
> *What did you give me to eat?*
> *I will give you dried figs or apricots*
> *For Me! What is this? Dried figs or apricots?*
> *Let me insert myself in the inner of the tooth and give me*
>   *his flesh . . .*
> *Out of the tooth I will suck his blood and from the gum*
>   *I will chew the marrow.*

For centuries, doctors and patients tried to smoke and burn the toothworm out. But for the most part the tiny critter remained a mystery, with a few notable exceptions. Dr. Cobaens, of the University of Copenhagen, saw a toothworm jump right out of his patient's mouth and grew the prize specimen in a bowl of water. Dr. Salmuth's toothworm was an inch and a half long—he said it looked like a cheese maggot. Dr. Shulz baited one with the gastric juice of a pig.

In 1733, Frenchman Pierre Fauchard, the father of scientific dentistry, declared that the toothworm didn't really exist, and it was never seen again.

Fight the Power

England's top doctors didn't much care for the poor, and did what they could to avoid them. But they didn't want poor people trying to get better on their own either, so when they published their long-awaited drug directory, the *London Pharmacopoeia* (1618), they made sure it was written in Latin. Happily, no one but the rich or well educated could read it.

Poor people weren't missing much. The book prescribed dried viper, crab's eyes, intestines of the earth (worms), frog spawn, sparrow's brains, penis of bull, rooster testicles, and unsalted butter. Still, it was the best medicine of its time, and a royal proclamation made the *Pharmacopoeia* the official catalog of all British pharmacies.

Born in 1616, Nicholas Culpeper trained for the ministry. He planned to elope, but his beloved was struck by a bolt of lightning and killed on her way to their secret rendezvous. His faith shaken, Culpeper became a pharmacist instead.

Culpeper worked with the poor. He wrote a readable version of the *Pharmacopoeia* in English. He did this so that, as he put it, doctors would stop "[w]riting prescriptions in mysterious Latin to hide [their] ignorance."

Elite doctors of the College of Physicians were aghast. They accused Culpeper of being a drunk, a degenerate, and a lecher, which he probably was. Thankfully for Culpeper, England's Star Chamber, a kangaroo court used to persecute unpopular defendants, had been recently abolished, so angry establishment doctors couldn't imprison or torture him, as they wanted.

Culpeper spent the rest of his life caring for "just folks" and showing them how to care for themselves.

Catching Babies

Delivering children, known as "catching babies," was women's work. In 1522, this fact was duly impressed upon a German physician named Wert, who, after dressing in women's clothes to watch a woman give birth, was burned at the stake.

With the eighteenth century came the rise of the "man-midwife." Man-midwives claimed that delivering babies required medical smarts and the use of tools, both beyond the reach of women. Women, in turn, claimed that the newly minted medical men just wanted to see them naked.

In 1748, twenty-eight eager male students, just returned from the War of Austrian Succession, swarmed the home of a woman giving birth. They caused a near riot, and the ruckus so rattled England's leading gynecologist, Dr. William Smellie, that he rushed the delivery and broke the baby's leg. Dr. Smellie later famously instructed his men to wear women's dresses so they'd appear less threatening and could hide their big tools under the material.

Watching men express so sudden a professional interest in the female anatomy, Elizabeth Nihill lashed out. The most prominent of the female midwives, she accused man-midwives of being "pudentists," butchers, and "broken barbers" and said one used to be a sausage stuffer. In pamphlets, she even mocked Smellie himself, suggesting he made house calls in a flowered calico nightgown with pink ribbons.

The men won the battle, but mothers and babies lost the war. Deliveries moved from women's bedrooms to filthy, germ-laden hospitals, and many more mothers and children died.

Blood Libel

> The days were passing and the Russian officials were waiting impatiently for his menstrual period to begin . . . If it didn't start soon they threatened to pump blood out of his penis with a machine . . .
>
> —BERNARD MALAMUD, *THE FIXER*

Though perhaps not sporting lizard tails, as some believed, Jewish men were thought to suffer excessively from painful hemorrhoids. According to Bernard de Gordon, one of the leading medieval physicians, this occurred for three reasons:

> First, because they are generally sedentary and therefore excessive melancholy humors collect; second, because they are usually in fear and anxiety . . . thirdly, it is the divine vengeance against them.

With the blossoming of the Renaissance, the sinister internal workings of the Jew received further scrutiny. When Spain's top inquisitor needed help persecuting Jews in 1632, court physician Juan de Quinones gladly stepped to the plate. He described in an influential pamphlet how "every month many of them suffer a flowing of blood from their posterior, as a perpetual sign of infamy and shame."

Mother's Milk

When it came to "nature's plan," rules were rules. According to child-rearing authority Dr. Jean Ballexserd, breastfeeding with the mother's own milk was mandatory. Any mother who didn't, no matter the reason, was an evil, heartless tramp. In 1762, Dr. Ballexserd wrote:

> [Y]ou have closed your eyes and heart to the call of nature. Look at the cat nursing its kittens . . . see the dog rejects food rather than abandon her puppies . . . O vain and pitiless woman, who can inhumanly reject a trust carried out by the fiercest of beasts.

Many doctors believed that milk from non-nursing mothers mixed with their blood and became poisonous, and that using foreign milk caused a mother's and baby's nerve fibers to vibrate at a different tension. Hiring a poorer, less educated "wet nurse," common among the upper classes, risked transmitting the nurse's "coarse disposition" to the infant: "Stupidity, anger, madness, and other affections are absorbed with the milk."

During one autopsy, unused milk was found on a mother's brain. This was only fair. When "the time for retribution had finally arrived," bad mothers would be "paid back in justice."

Heartless mothers faced another hurdle—disposing of the untapped milk in their breasts. They'd use leeches or little dogs, but try as they might, they could never trick their own bodies or fool a vengeful Mother Nature.

The Secret

In the sixteenth century, Peter Chamberlen invented a new type of forceps. The design was ingenious enough to create a mini-revolution in obstetrics. That revolution never happened, because Chamberlen kept it a secret.

In one publication, Chamberlen chastised other doctors for using "hooks" that killed both baby and mother. He bragged of his own lifesaving device, then voiced regret that he couldn't say more—business was business.

The Chamberlen family kept the secret for two hundred years. The forceps were hidden in plain sight, often in a striking box decorated to look like a pirate's booty. Secured with locks, the treasure chest was so heavy it had to be carried into the house by two strong men.

During delivery, bedroom doors would be closed, curtains drawn, and the mother blindfolded. Henchmen would pound hammers and ring bells, anything to distract, confound, and deepen the mystery.

In the late 1600s, Peter's son Hugh, who may have needed money to bankroll a bogus real estate scheme, decided that selling the secret was more profitable than keeping it. His deal with the French, however, fell through when a mother and her baby died during a sales demonstration. In 1693, Hugh finally sold the secret to the Dutch. He gave them only half of the forceps, which, not knowing any better, they happily accepted.

Along with struggling newborns and their mothers, the secret of the forceps died with the Chamberlens. But in 1813, a maid cleaning their old house found, hidden under the floorboards, a trapdoor concealing a secret chamber. In the chamber was a box. In the box were love letters, and Peter Chamberlen's original forceps, from the sixteenth century. No longer a secret, they are now exhibited at London's Royal College of Obstetricians and Gynecologists.

Through the Looking Glass

Galileo knew a good thing when he saw it. Learning of a new type of looking glass around 1610, he got one for himself and soon began making his own. With his new "telescope," he saw distant stars and planets, and modern astronomy was born. The microscope, invented at exactly the same time, didn't make quite the same splash. It was ignored.

Seventy-five years later, Italian doctor Giovanni Bonomo watched peasants scratch at bites. He pricked "an itchy person" and with his microscope saw for the first time "a living creature, in shape resembling a tortoise . . . of nimble motion, with six feet, a sharp head [and] two little horns at the end." It was the scabies mite, which he realized jumped from person to person.

This wonderment, not to mention Leeuwenhoek's discovery of the amoeba, proved of little interest to the medical establishment, which had neither the time nor the inclination to look at things too closely. In 1692, noted scientist Robert Hooke wrote that the microscope was:

> [n]ow reduced almost to a single votary, which is Mr. Leeuwenhoek; besides whom, I hear of none that make any other use of that instrument, but for diversion and pastime.

Thomas Sydenham, "the English Hippocrates," dismissed the microscope, and much of anatomy to boot. Nature, he felt, would never betray her secrets, and anyway, medical inquiry should be limited to "the outer husk of things." Jan Swammerdam, himself a pioneer in microscopy, thought that looking at God's handiwork too closely would lessen mankind's sense of wonder. He quit altogether.

Historians later claimed that early microscopes were too crude to be of great use. This hypothesis was tested in

1989, when nine of Leeuwenhoek's original microscopes were tested. They were found equal to or better than today's beginners' models.

Warm Beer

In 1641, English medical experts published *Warm Beer: Or a Treatise Wherein It Is Declared That Beer So Qualified Is Far More Wholesome Than That Which Is Drank Cold*:

> *Some will say that cold beer is very pleasant to one that is thirsty: I answer it is true, but pleasant things are for the most part very dangerous . . .*
>
> *How many of you have known and heard of [people], who by drinking a cup of cold beer in extream thirst, have . . . killed themselves? What's more pleasant than for one that hath gone up a hill in summer-time and is exceeding hot, to sit down and open his breast . . . and yet how dangerous.*

In Marfield, England, a Mr. Hammerman dropped dead after a cold brew, and cold beer almost killed Mr. Clark's wife, who already had cancer of the matrix. A famous gentleman in Italy, whom the book's authors preferred remain anonymous, became so sick he couldn't hold his water—it "cometh from him without his Knowledge."

The danger of cold beer could be inferred from Hippocrates, Galen, and Aristotle's *Fourth Book of Meteors*.

Cold beer hardened meat and prevented it from properly boiling in the stomach. This was easily proved, because "[T]hem that drunk much cold Beer after Dinner or Supper . . . they will vomit up the same Meat again . . ."

Cold beer also caused "imbecility of the joints" and inflamed the meninges of the brain to "an exquisite Phrenzy."

Warm beer, on the other hand:

> [h]elps the stomach, and by that means the head, and by that means the liver, and by that means the bowels, and by that means the spleen, and by that means the Kidneys and Bladder, and by that means the Matrix in the Women . . . and by that means preserves old age and consequently preserves life.

Warm Beer achieved great popularity with both doctors and beer drinkers and came out in four editions over the span of a century.

It's Witchcraft

With the Enlightenment came a more nuanced understanding of hellish evil. Demonic possession and its symptom, witchcraft, became a diagnosable medical disorder, like the measles.

A doctor would shave a suspect and examine his skin for "devil's stigmata." If evidence of devilry was found, he'd supervise the torture.

In 1628, Johannes Junius described his plight in a letter smuggled to his daughter:

Many hundred thousand good-nights, dearly beloved daughter Veronica. Innocent have I come into prison, innocent have I been tortured, innocent must I die . . . When I was the first time put to the torture, Dr. Braun, Dr. Kotzendorffer, and two strange doctors were there.

After being burned and tortured with thumbscrews, leg screws, and the strappado, Junius admitted he had changed his name to Krix, was seduced by a bleating goat, and rode a black dog to a witch dance.

Many doctors steered clear of the hysteria, and a few brave souls went public with their opposition. Professor Wagstaffe of Oxford disputed that witches could stand in the air and run up walls without using their hands, but he was accused of "bibbing . . . strong and high-tasted liquors." Dr. Weyer of Denmark, said to have coined the term *mental illness*, courageously argued against demonic possession, though he himself had discovered that 7,405,926 evil spirits existed in the world, supervised by seventy-two Princes of Darkness.

Some medical luminaries actively encouraged witch hunting. When a team of doctors found a patient to be suffering from epilepsy, Ambroise Paré, the father of modern surgery, did his own exam and declared him possessed by evil spirits. Paré also described, in precise anatomical detail, how devils entered the body and showered it with torments. On hearing King Charles IX laugh at the antics of a court magician, Paré whispered in the king's ear, "Thou shall not suffer a witch to live."

Sir Thomas Browne, England's celebrated debunker of medical myth, was another famous witch hunter. Knighted for his advocacy of "The New Learning," Browne testified at the trial of Rose Cullender and Amy Duny, accused of blinding children and making them vomit nails. His expert testimony convinced a jury that the girls were witches, just like the ones in Denmark.

The Tarantella

A man poor was taken ill in the street . . . The people at the sight cried out—play-play the tarantella . . . The first two bars, the man began to move accordingly, and got up, quick as lightning . . . but as I'd not as yet learn'd the whole tune, I left off playing . . . But the instant I left off playing the man fell down and cried out very loud . . . [he] scraped the earth with his hands . . . in miserable agonies. I was frightened out of my wits.

—NAPLES TOURIST STEPHANO STORACE

During the summer, when the sun grew too hot, exhausted Italian peasants would stream from the fields and begin dancing in the streets. Saying they were bitten by a spider, they'd dance all day and all night, and they'd eat and drink, and laugh and flirt. News of the spider bites would spread, and soon the entire town would be stricken.

To cope with the epidemic, village officials would have, in advance, hired entertainers and rented out halls. After a few days, exhausted spider bite victims would declare themselves "cured," until next summer.

In the fifteenth century, an amused Giovanni Pontano wrote of tarantism:

> [T]he people of Apulia are extremely happy, because while other men have no excuses for their follies, the people of Apulia always have a ready one, the tarantula, to which they attribute their insane desires.

Doctors weren't quite so skeptical. In the 1640s, Athanasius Kircher, who invented the megaphone, thought the armadillo a cross between a turtle and a porcupine, and designed a cat piano (its strings slammed down on a cat's tail), journeyed to Rome and captured tarantulas in a custom-designed glass vial. Isolating their toxin, Kircher discovered that tarantism, left untreated, could create a desire to fondle things that were purple, and that some spider bite victims, thinking they were ducks, would grab reeds from a pond and dive underwater. Renaissance man Epiphanio Ferdinando, whose spider-bit cousin died when the music stopped, wrote that virgins would rip off their clothes and that men bearing swords would jump into the sea and wallow in the mud like pigs.

Zithers were long regarded as a near-universal antidote, but Kircher claimed that different instruments affected different people in different ways. A teenage girl, for example, might groan at a slow beat, while grown men might weep at the sound of a badly tuned violin.

Kircher also felt that certain types of music, played just right, exerted a magnetic effect on a spider's poison, drawing it out of the body. English doctor Walter Charleton argued that music, combined with dancing, sped up fermentation in the brain, expelling the poison faster.

The real mystery was what caused tarantism's bizarre symptoms in the first place. Some doctors believed that spiders, long repressed, played out their own deepest longings and desires by transferring them, via spider bite, to their innocent victims.

The Royal Flush

In ancient Egypt, the clyster (enema) was said to have been invented by the graceful ibis bird, who, with its long, tapered beak, could supposedly administer an enema to itself. In Renaissance France, the taking of enemas reached similarly fashionable status among the upper crust.

Louis XI of France gave enemas to his pet dogs, and Louis XIII had 212 enemas in one year, along with 215 vomiting sessions and 47 bloodlettings. Louis XIV, the King of Clyster, had more than two thousand enemas, sometimes four times in a day. They apparently worked— he lasted seventy-two years on the throne, successfully prosecuted the War of Spanish Succession, and eliminated the last vestiges of feudalism.

In the sumptuous palace at Versailles, palace maids, prohibited in writing from sneaking enemas themselves, would mix, tint, and fragrance enema solutions for their masters, whose closets bulged with custom-fitted enema tubes made of silver. The tubes were packaged in velvet sacks with drawstrings. To keep minds sharp and complexions clear, royals were encouraged to have at least one enema a day, with an after-dinner enema as a nightcap.

To receive an enema, His Royal Highness would have been asked to "bend one leg forward and expose all that is required, without shame or false modesty." Administering an enema could be like conducting an orchestra:

> [R]everentially placing one knee on the floor, he will guide his instrument in his left hand, without rushing or flailing, and with his right hand, he will push the plunger "amoroso" [lovingly], with control and without any jerks, "pianissimo," very, very slowly.

The enema was naturally the subject of palace intrigue. Louis XIV so feared being poisoned that he created a special bureau to hunt down enema assassins, and chaired high-level meetings with his ass bared. To enhance her prestige at court, the infamous Madame de Maintenon made Ninon de Lenclos give her an enema in public.

Outside the palace, many pharmacists specialized in giving enemas. They were nicknamed *limonadiers du posterieur*, or lemonade makers of the ass. As a selling point, they hung signs depicting rectal nozzles and plungers. In 1668, however, a physician named De Graff published his fiery *De Clysterbus*, which for the first time made self-administered enemas, with copper pipes, accessible to the masses.

But the shit of the upper class still stank, and among the well-to-do it fell to servants to clean up the mess. Popular revulsion may well have foretold the coming revolution:

I've had quite enough of the stink that I smell
I want to go out. Let the Docs go to hell
They condemn Madam to the chair with the hole
Forcing me yet again to clean out the bowl.

The Bezoar

At the fabled Library of Alexandria, ancient Greek doctors are said to have butchered up to six hundred people. Historian Celsus, a big admirer, gushed, "Herophilus and Erasistratis did anatomical studies in the best possible way, laying open men while alive."

But medicine rarely used live people as guinea pigs. In 1575, Ambroise Paré, the greatest surgeon of the Renaissance, broke that taboo.

The bezoar was a hard, stonelike clump occasionally found in the stomachs of people or animals, chock-full of magical powers. Bezoars were believed to be a universal cure for poisoning, so powerful that, as an insurance policy against palace intrigue, kings had them embedded in their drinking mugs.

Presented with a rare bezoar in 1575, King Charles IX of France exulted over its lifesaving properties. Paré was skeptical but told the king that actually testing the bezoar would be "an easy matter."

Having stolen two silver dishes, one of the palace cooks was about to be executed. At Paré's suggestion, he was promised a pardon if he took poison and survived the ordeal. The cook swallowed the bezoar, then the poison. With three of the king's archers, Paré went down to the dungeon to watch.

Paré took copious notes. First the subject writhed in pain, and then he screamed that he was on fire. Paré then wrote:

> I found the poor cook on all fours, going like an animal, his tongue out of his mouth, his face and his eyes flaming red . . . he died miserably, crying out he would better have died in the gallows; he lived seven hours.

Asked years later about the incident, a satisfied Paré expressed no regret.

Out of Thin Air

The Egyptians thought crocodiles were formed from the mud of the Nile, and Leonardo da Vinci believed frogs, bats, and fish came from the stars. But spontaneous generation, the theory that inanimate objects give rise to living creatures, seemed ripe for reexamination.

Attention turned to the lemming, which, according to most thinking people of the sixteenth century, was created out of thin air. Olaus Magnus, the great Swedish patriot and scholar, thought lemmings were carried by the wind and dumped on people's heads, especially near Helsinki. Zeigler of Strasbourg proposed that lemmings fell out of the sky during storms and died when grass grew in the spring.

In the seventeenth century, Danish physician Ole Worm dared oppose prevailing orthodoxy. Discovering that lemmings had testicles, he proposed that little lemmings really came from big lemmings.

At about the same time, Dr. Jan Baptist van Helmont, the founder of pneumatic chemistry, took dirty underwear from "an unclean woman," put them in a jar of wheat for twenty-one days, and watched as mice jumped out. To him, this reaffirmed once and for all the truth of spontaneous generation.

Several years later, Rudbeckius the Younger refined the theory. The wife of a Swedish merchant cooked her husband duck-egg pancakes. Opening the box eight days later, the merchant saw several frogs leap out, but no pancakes. Rejecting the conventional explanation—that the frogs had arisen spontaneously from the pancakes—Rudbeckius proposed that when the mother duck first laid her eggs, before they went into the batter, she had been munching on small frogs. The "essence" of these frogs, their "life seeds," had then been transferred to the duck eggs. When the box was opened, they hatched as frogs.

· CHAPTER FOUR ·

The Age of
Heroic Medicine

Whenever a doctor cannot do good,
he must be kept from doing harm.
—Hippocrates

L IKE firemen rushing into a burning building, doctors during medicine's "Heroic Era" considered themselves heroes. That they may have been fanning the flames instead of putting them out didn't trouble them in the slightest. In fact, so busy were heroic doctors burning, drowning, spinning, and electrifying their patients, they barely seemed to notice at all.

During medicine's Heroic Era, running roughly from 1780 to 1850,* bloodletting and the purging of ill-defined "humors" from the body reached their peak. With new techniques to try and new technologies to play with, it was bombs away on illness and disease, and patients as well.

---

* Heroic medicine continued to be practiced after 1850. A few post-1850 examples are included here.

Medical thinking wasn't much better. Thomas Jefferson called medicine a "pious fraud." In an 1847 essay published in the *New York Journal of Medicine*, Dr. N. S. Davis, a founding member of the American Medical Association, wrote:

> [T]he contradictory nature, and the absurdity of these systems or pretended systems of medical philosophy, that has caused medical science to be regarded by very many as so vague and uncertain as to be little better than a blind system of guessing.

Germ theory and anesthesia gave medicine its revolution, finally, but doctors' simple failure to keep track of their successes and failures, and learn from them, doomed generations more. Meanwhile, the rise of widely read medical journals like England's prestigious the *Lancet*, named after the beloved bloodletting tool, lent legitimacy to even the most outlandish of medical schemes.

During the Heroic Era, medical thought and practice may well have reached their lowest point since the time of Hippocrates, more than two thousand years earlier.

Diseases of the Mind

Treasurer of the Mint and signer of the Declaration of Independence and founding father Benjamin Rush was America's best-known doctor. According to one historian, he "shed more blood than any general in history."

Rush's reign of error began during Philadelphia's yellow fever epidemic of 1793, when he insisted the deadly outbreak was caused by the smell of coffee beans rotting along the Delaware River. He was sure the fever wasn't contagious, even after his sister and three assistants died from it. Dr. Rush blistered his patients, wrapped them in vinegar, doused them with cold water, and gave them "artificial diarrhea." Most of all he bled them—in Mr. D. T.'s case, a total of four gallons.

In 1773, Rush published an abolitionist pamphlet and became a leader of the anti-slavery movement. In 1776, he bought a slave. In 1884, he held on to the slave when he joined the Pennsylvania Abolition Society.

Still considered the father of American psychiatry, Rush discovered that mental illness was caused by poor circulation of blood to the ventricles of the brain, along with bad weather, blood transfusion from animal to human, worms, and "sympathy" between the brain and hemorrhoids.

Environmental stressors also triggered mental illness, according to Rush. To this he devoted much of *Medical Inquiries and Observations, Upon the Diseases of the Mind* (1812), which became the leading psychiatric textbook for almost fifty years.

Citing Baron Humboldt's well-known proposition, "[It] is a rare thing . . . to see a Russian peasant angry," Rush claimed that the hurly-burly of everyday life caused psychotic episodes. He wrote of overly brainy patients driven mad by thoughts of perpetual motion and alchemy, and of an actor hissed off the stage, and of the formerly beautiful Lady Montague, who went insane when she looked in a mirror for the first time in eleven years.

Dr. Rush is best remembered for his advocacy of humane treatment of the mentally ill. At his Pennsylvania Hospital, he threw away the chains and insisted that patients be treated like family; sometimes he brought them fruitcake. He also made them suffer, for their own good. Madmen, he knew, experienced great tranquillity when they were about to drown.

Dr. Rush poured acid on his patient's backs and cut them with knives. He kept their wounds open for "months or years," to facilitate "permanent discharge from the neighborhood of the brain." Hearing that the wild elephants of interior India were tamed through starvation, he recommended doing the same for the sick and suffering.

News of Rush's therapeutics spread. Dr. Gregory of Scotland yoked several madmen to a farmer's plow. A European doctor, treating a patient who thought he was a plant, watered him with a teapot of urine. Dr. Rush wrote approvingly of a colleague who put an artificial snake in his patient's snuffbox, and of another who complimented his patient's drawing of a cabbage, knowing all the while it was meant to be a beautiful flower.

To some, Rush's greatest contribution to the field of psychiatry lay in the area of "swinging." Strapped to chairs hung from the ceiling by a length of chain, his patients were spun like tops for hours on end. "I have called it a Gyrator," he announced. Dr. Rush also developed the widely used "tranquilizer chair," which, with a hole in the bottom for defecation, blinded, gagged, and froze a patient in place for hours or days. A *Wheel of Fortune*–like "centrifugal board," to which patients were to be strapped and spun, was designed by Rush but apparently never used at the hospital.

Dr. Rush's likeness still adorns the seal of the American Psychiatric Association.

Laughing Gas

Joseph Priestley discovered laughing gas (nitrous oxide) in 1772. His other big invention was soda water. For years, nitrous oxide was snorted through the upturned noses of idling English gentry—they used it to party.

In 1798, British scientist Sir Humphry Davy, never averse to a good time himself, began his famous experiments with nitrous oxide, tripping with friends. One friend compared it to champagne; another said, "I felt like the sound of a harp." Davy came to really enjoy his work. "I noticed that I have breathed it [the gas] very often," he wrote. Also, "[A] desire to breathe the gas was especially awakened when I caught sight of [a] person breath[ing] in deeply." Davy was hooked.

In 1799, Davy embarked on his newest project: to discover the cure for a hangover. According to Davy's work log, he inhaled laughing gas, drank alcohol, threw up, and felt much better. Three days later, Davy stepped into a boxful of laughing gas, sealed airtight. Notes reveal that he greatly enjoyed the experience. Davy wrote in his diary that nitrous oxide "[a]ppears capable of totally destroying physical pain" and recommended its use during surgery.

He then forgot all about it. Decades of surgical patients went without.

In 1844, dentist Horace Wells attended a nitrous oxide show, where, according to the circus poster, forty gallons were to be dumped on the audience. "Eight strong men" were in attendance, just in case. Wells, stoned, realized the gas's value as an anesthetic. But he bungled a big hospital demonstration, and jeering medical students shouted "Humbug! Humbug!" as he ran from the room. Wells later had a nervous breakdown and switched to selling canaries. In 1848, he sprayed sulfuric acid on two prostitutes and then killed himself, though not before taking nitrous oxide to deaden the pain.

Ether was traditionally used for "ether frolics." It took three hundred years for doctors to finally wake up.

In 1846, W. T. G. Morton, an unlicensed dentist and con man, borrowed (or stole) the idea of using ether as an anesthetic. He colored the invisible gas, renamed it "Letheon," and pretended to have discovered something new, hoping to cash in on a new patent. But doctors caught on and within weeks operated painlessly all around the world.

In 1847, Sir James Simpson, personal doctor to Queen Victoria, began his own historic work with chloroform. Of an early "experiment," one participant wrote, "[I saw] professor [Simpson] in the midst of a group of girls . . . I came round to a consciousness . . . of the professor flitting about in great glee."

Someone who witnessed Simpson's more advanced research wrote:

I'm beginning to fly!" she shouted. "I'm an angel, oh, I'm an angel! Dr. Duncan, Mrs. Simpson, the professor him-

self ... shouted loudly and roared with laughter ... Only the naval officer sat looking on, puzzled and aloof ... Then he began to crow like a cock ...

Dr. Simpson leaped up from his chair and stood on his head ... he fell with a crash onto the floor ... "This is far stronger and better than ether," he said.

Spin Doctors

Dizzy doctors spun patients in human centrifuges, hoping to separate the bad humors in their heads.

In *Zoonomia* (1801), Dr. Erasmus Darwin, grandfather to Charles and later to bring a strand of vermicelli to life in a glass vial,* introduced the world to his "rotative couch." The sofa, though not well upholstered, was designed to let patients sleep while they spun.

Darwin's couch inspired mental health pioneers like Joseph Mason Cox, of Fishponds Private Lunatic Asylum, and William Halloran. They created their own "circulating swings," which spun patients a hundred times a minute. Halloran relished his role as a disciplinarian. "[S]ince the commencement of its use, I have never been at a loss for a direct mode of establishing a supreme authority over the most turbulent and unruly."

A few years later, Dr. Ernst Horn of Berlin tucked hundreds of his patients into a rotating bed thirteen feet

---

* Darwin's "reanimation" of the vermicelli was one inspiration for Mary Shelley's *Frankenstein, or the Modern Prometheus.*

in diameter, spun by a crankshaft and controlled by ropes hung from the ceiling. Horn's therapy lasted sixty to ninety seconds; the bed spun at up to 120 revolutions a minute. Later a general in the Prussian army, Horn considered the centrifuge a part of his "Order Therapy." "[T]he more lively his intimidation toward the apparatus . . . the more charitable the effects of the therapy," he wrote.

Horn was dismissed from his post for suffocating a female patient in a burlap sack, but that did little to quell enthusiasm for his elaborate spin cycle, which was widely employed in several countries. A working model still exists at the former Waldau Insane Asylum in Bern, Switzerland.

The Changing of the Peas

[A]n issue is an artificial sore . . . [It is] formed in various ways, such as burning with a red hot iron, by caustic etc., but the most general and indeed common form in popular use is the pea issue . . . In the course of a few days the irritation occasioned by the peas causes the discharge of matter; the peas are generally changed every day or two days.

—*A DICTIONARY OF DOMESTIC MEDICINE AND
HOUSEHOLD SURGERY* (1852)

Adding insult to injury, "counter-irritation" let doctors treat, or pretend to treat, illnesses they knew nothing about. It was the subject of long volumes and countless meetings and congresses, but doctors never quite figured out how or why counter-irritation worked, which it didn't.

Dr. Dendy wrote of antistasis, metastasis, and transmigration; Hunter wrote of revulsion, repulsion, and derivation; and Boyle wrote of an organized deposition of coaguable lymph. In opening remarks to a meeting of the Royal Society of Medicine in 1909, a jocular President Brown compared the theory behind counter-irritation to the case of the French criminal who, being tortured on the wheel, noted that the second turn wasn't half as bad as the first.

To "irritate" a patient was to wound him. In *The Theory and Practice of Counter-Irritation* (1895), Dr. Gilles referred to the doctors of the past, who had stuck their patients' inflamed limbs into giant anthills. But Dr. Gilles now recommended use of the fire iron and also blistering with a "caustic" (strong acid). The key to using a caustic, he wrote, was to produce a blister twice as big as the ultimate wound. Quicklime was good.

Doctors would create several small blisters or one large blister, though in 1824 Dr. Abernathy, writing for the *Lancet*, said a blister a foot square was probably too big. Lecturing at a medical convention in 1909, Dr. Soper spoke of patient "fear and trepidation" and fondly recounted applying an eight-by-four-inch blister for pneumonia.

More common than blistering was the making of an "issue." Wielding a scalpel-like *seton*, doctors would cut into their patients with a "sawing motion." Foreign objects, usually dried peas or beans, would then be inserted into the gash to promote proper infection and oozing. A doctor would reopen the wound, often every day, for weeks or months afterward, to make sure it didn't heal.

Dr. Jonathan Toogood, senior surgeon at Britain's Bridgewater Hospital, wrote of treating a twenty-year-old woman with a hernia:

> I made an issue on either side of the curvature, large enough to contain, in each, forty small horsebeans, which was . . . kept open for upwards of two years . . . she was not entirely confined to the house . . . by day, she rested on a small-four-wheeled carriage . . . she could work, draw and amuse herself.

Beat the Clock

[T]he operator's success will be in direct ratio to his quickness.

—FLORENCE NIGHTINGALE, *NOTES ON NURSING*

In the nineteenth century, surgeons would have students bring stopwatches to their surgeries. Speed was a matter of pride, but the race wasn't just for show. Every passing second increased the risk of shock and infection and, in an age before anesthesia, bloodcurdling screams. A good surgeon could amputate a leg in ninety seconds.

Robert Liston was one of the greats. According to Gordon, his biographer:

He sprung across the blood-stained boards . . . like a duelist, calling "Time me gentlemen, time me!" to students craning with stopwatches . . . Everyone swore that the first flash of his knife was followed so swiftly by the rasp of the saw on bone that that sight and sound seemed simultaneous . . . He would clasp the bloody knife between his teeth.

In just such fashion, Liston took four minutes to remove a forty-five-pound scrotal tumor, which before the operation had to be carried by the patient in a wheelbarrow.

With speed came risk. Liston once cut off a patient's testicles while amputating his limb. After being told that a pulsating tumor in a boy's neck could be a dangerous aneurysm, the impetuous Liston grabbed a knife from his pocket and cut anyway. The boy fell dead on the spot. The boy's severed artery is now on display at University College Hospital in London.

Dr. Liston's most notorious case occurred while amputating a leg. His patient eventually died of gangrene, not unusual in the days before sterile surgery. But during the procedure Liston also cut off the fingers of his assistant, who proceeded to die of gangrene as well. And, according to Gordon, Liston slashed through the clothing of a spectator, who, terrified over seeing his own skin punctured, dropped dead from fright.

Suckers

It is now found more economical to feed the leeches on cows. The heavy, dull animal, haggard, frightened yet resigned to its fate, bears the onslaught of the leeches . . . with a sort of stupid surprise.

—BRITISH MEDICAL JOURNAL (1863)

The French went through forty million leeches a year, and many doctors feared for their extinction. To meet the demand, women waded into leech-infested ponds and let them cling to their legs.

Professional leechers did hospital rounds, and local drugstores sold leeches by the barrel. Sir Samuel Romilly, the famous British legal reformer, kept his pair of leeches as pets. He fed them every day and gave them names.

A doctor would tie a leech to some silk thread and lower it down his patient's throat. When the leech became heavy with blood, he'd reel it in like a fish. To bleed a man's testicles, doctors often applied, over the course of several days, a hundred or more leeches. Astley Cooper, the great English surgeon, reported that many men, afraid that leech bites to the groin were a turnoff, chose the lancet instead.

Doctors commonly applied leeches to the anus. This had to be done with caution, to prevent patients from going into contraction or spasm. In 1818, Dr. William Brown suggested using a chair with a hole in the bottom and a bucket underneath. He applied his leeches with a long-necked bottle. Dr. Osborne recommended use of a grooved rod, firmly thrust upward. His had a decorated leather handle.

To relieve uterine disease, sexual excitement, and "exasperation" in general, leading textbooks and journals strongly advocated the application of leeches to the vagina. In England, "men of high standing" had their wives leeched every two weeks.

Again, doctors had to be careful. As one purveyor of "artificial leeches" happily noted in his medical advertisement, real leeches could "crawl into cavities or passages result[ing] sometimes in very annoying accidents."

According to *A Handbook of Uterine Therapeutics and of Diseases of Women* (1868), in pregnant women "a jet of blood" would spurt when a leech grabbed on to the wrong vein, and even the smoothest procedures risked blood clots or "flooding," and sometimes a miscarriage.

The textbook also dealt, at some length, with leeches getting "lost" inside a woman. Two leeches out of six, for example, might go missing. In such cases, it was strongly recommended that the doctor not tell his patient, lest she go into needless "hysterics." Women did tend to be highstrung, the doctor reported, but eventually, "the leech is sure to find its way out."

Dead Weight

Doctors always craved dead bodies. During the Renaissance, Professor Monro supposedly gave 124 lectures using the same opened corpse, while Richter of Germany taught anatomy using a large turnip. All Professor Rollfink of Germany did in 1629 was request a dead body, and he was stoned to death by an angry mob. Before their execution, his killers asked only that they not be "Rollfinked" themselves.

English doctors hungered for bodies, the fresher the better, and didn't much care how they got them. With body snatching big business, British lawmakers passed the Murder Act in 1751. As if execution weren't enough, the law prohibited the decent burial of a hanged murderer's body. Not coincidentally, it also ensured fast delivery of corpses to doctors and anatomists, who stood waiting by their tables.

At hangings, grieving relatives and "surgeons' agents" often fought for possession of the body, detracting from the otherwise festive atmosphere. Brokers and bidders would jostle for position and maybe even haggle with the condemned man himself, who could trade his body-to-be for decent clothes to be hanged in.

Family members would lurk nearby, hoping to snatch the body for quick burial. Because hanged men often died slowly of asphyxiation, friends and family would grab hold of the convulsing body and swing along with it, hoping gravity would do what the hangman hadn't.

Writer Samuel Richardson described such a scene, in 1741:

> As soon as the poor creatures were half-dead, I was much surprised ... to see the populace fall to hauling and pulling the carcasses with so much earnestness, as to occasion several warm recounters and broken heads. They were friends of the persons executed ... and some persons sent by private surgeons to obtain bodies for dis-

section. The contests between these were fierce and bloody, and frightful to look at.

Once in a while, a hanged body woke up. Anne Green was seduced by her master's grandson and in 1651 gave birth to his boy. Green claimed, and medical evidence indicated, that the boy was born dead, but she was convicted of murder anyway. At Green's request, friends jumped on her hanging body and struck repeated blows to her head. Hours later, however, while on the dissection table, she began to breathe. The anatomists took pity and revived her with whiskey. Fully recovered, Green was granted a pardon.

Let the River Run

Hemophilia (literally, the "love of blood") was treated cruelly, if at all.

In ancient times the disease was essentially ignored, though a Spanish village of "bleeders" was written about in the tenth century. In 1539, a barber is said to have died after scratching his nose with a pair of scissors.

Attention was paid in the eighteenth century. Medical records reveal that members of the tragic Mampel family died from cutting their lip, falling off a chair, jumping over a tree trunk, hitting their mouth against a door, being run over by a wagon, tripping over stones while drunk, and falling with wood in their mouth. A hemophiliac from another family supposedly outlived everyone by confining himself to a chair for thirty years.

By the 1830s, experts knew hemophilia was caused by the failure of blood to coagulate, but doctors overlooked this simple fact for seventy-five years. In the meantime hemophilic patients were often talked into having quality-of-life procedures—knee surgery, for example—with sometimes fatal consequences.

Many doctors thought it best to let patients bleed themselves dry. "One of the most satisfactory ways of treating bleeders is to leave them to themselves," said one leading medical journal. A hospital did just that to a three-year-old boy who cut his tongue. "[B]lood gushed from his mouth . . . the bleeding continued for seven days until he was like a wax doll in appearance . . ." The boy died.

Other doctors were more aggressive. At the famed Salpetriere Hospital in Paris, doctors applied twenty-five leeches to the anus of a known hemophiliac; he died too.*

Through it all, doctors blamed hemophiliacs themselves, calling them reckless and irresponsible. Hemophiliacs were attacked for, among other things, walking, going to work, and living in a house without round furniture.

In 1898, a letter in the *British Medical Journal* bemoaned hemophiliacs' "constant mental peculiarities" and "unwillingness to tell the truth." One disgusted doctor treated a teenage boy with blood running from his mouth. Embarrassed, the boy denied being a "bleeder." When his mother said otherwise, the boy threw a tantrum. With his

---

* On numerous occasions, evil genius Rasputin stanched the bleeding of hemophiliac Alexei Nikolaevich, heir apparent to the Russian throne, by telephone. Rasputin's soothing voice apparently lowered the boy's blood pressure.

dying breath, the boy insisted he bled no more than any other boy. The letter writer declared the dead boy, and others like him, a menace. Another doctor responded to the letter. He personally knew of two hemophiliac boys who were also liars. A third was so nasty that his parents, fed up, hoped he would die.

Show Business

Kangaroo boxing had its charms, and, in eighteenth-century London, demonstrations of static electricity and learned pigs surely had their followers. But nothing equaled the spectacle of watching a freshly delivered human body demolished by a master anatomist. Done right, public anatomy was entertaining, enlightening, and bloody as a cockfight.

Human dissections were held in public halls or bars, or, for private parties, in a person's living room. The Blew Boar tavern, located near the hanging ground, advertised its anatomical events in the newspaper. At festival time, colleges and museums sponsored their own human dissections, with free admission for all. Crowds could be raucous, and if things got dull, a live animal could be slaughtered.

Anatomical demonstrations in Berlin were the hottest ticket in town, and the Paris Academy sold out. Berlin charged more for the rare female body, and still more if the woman was pregnant.

At the opening of a theater in Dresden, an overflow crowd came to see a woman who had been decapitated. The ladies were allowed to touch the headless corpse, and the performance was held over for another day. Things may have peaked in 1780, when, in Jena, Germany, fashionable Duchess von Weimar attended the dissection of several dead children.

Live Wires

The shock of giant electric eels (*gymnoti*) supposedly jolted people into wellness. In 1800, Alexander von Humboldt, whom Charles Darwin described as "the greatest scientific traveler who ever lived," went stalking them with French physician Aimé Bonpland. They cornered their prey in a South American lake:

> [R]esembling large aquatic serpents [they] swim on the surface of the water, and crowd under the bellies of the horses and mules . . .
>
> [T]he eels, stunned by the noise, defend themselves by the repeated discharge of their electric batteries . . . Several horses sink . . . stunned by the force and frequency of the shocks, they disappear under the water . . .
>
> The eel being five feet long, and pressing itself against the belly of the horses . . . attacks at once the heart, the intestines, and the . . . abdominal nerve . . . By degrees . . . the wearied gymnoti (eels) dispersed. They require a long rest . . . to repair the galvanic force which they have lost.

> *The mules and horses appear less frightened . . . The*
> *gymnoti approach timidly the edge of the marsh, where*
> *they are taken by means of small harpoons . . . In a few*
> *minutes we had five large eels.*

Humboldt and Bonpland performed medical experiments with their eels but ultimately couldn't compete with newly developed electric batteries.

Burking

to Burke: verb (used with object), burked, burking
to murder, as by suffocation

To get at a grave, "resurrection men" would dig down vertically, tunnel horizontally, and then snatch the body, using a wooden shovel to dampen the noise. Cemeteries fought back, employing guards, watchtowers, and even land mines. With bodies at a premium, British doctors and hospitals encouraged the thefts, from a respectable distance.

Body snatching was dirty and difficult, so William Burke and William Hare decided to skip the middleman and kill people themselves. In Dr. Robert Knox, an anatomy teacher at the Royal College of Surgeons in Scotland, they found an enthusiastic buyer, known for his no-questions-asked policy.

"Daft Jamie," a mentally disabled teen, went missing. His mother went looking for him, even as Dr. Knox laid his body on the table. When a few of Knox's students thought they recognized Daft Jamie, Knox cut his face off, and then his hands and feet.

> Burke's the butcher, Hare's the thief,
> Knox, the boy who buys the beef.

Burke and Hare were implicated in at least sixteen murders, and Burke was condemned to death. Dr. Knox was never censured in the slightest and continued to teach and dissect. Outrage over this and other incidents led to the passage of the Anatomy Act (1832), which greatly expanded the supply of corpses.

Burke was hanged. The professor who dissected his body took a quill pen and inscribed, "This is written with the blood of Wm. Burke." Burke's skin was made into a pocketbook, now on exhibit at Edinburgh Medical College.

Talk Therapy

Demosthenes, the Greek orator, overcame his stutter by talking with pebbles in his mouth. In the tenth century, influential Rhazes blistered and burned his stammering patients. Once, to rid the brain of excess humidity, he put his patient's head in a plaster cast. Mercurialis of the sixteenth century recommended not taking a bath and eating fewer pastries.

In 1830, French physician Hervez Chegoin proposed that people stuttered because their tongues were too short or not stuck in their mouths right. Only "mechanical means" could correct the problem, and he commenced with his surgeries. Sometimes a person didn't even need a stutter to be operated on. An 1842 journal article notes that of forty-two surgeries for stuttering, two were done on people who didn't stutter at all. They had poor diction.

The leader in the field, Johann Dieffenbach, once said, "I was first led to the idea of operating on the tongue by hearing a stuttering person request me to cure him of a squint . . ."

In an 1841 address to the Institute of France, Dieffenbach described one of his first surgeries, on thirteen-year-old Fredic Daenau. He had the boy stick out his tongue and grabbed it with a double hook. Then he cut out a wedge-shaped piece, about eighteen millimeters. "The quantity of blood lost was considerable," he wrote. A journal article later reported, "The stutterer can be temporarily cured for a few days by making a slight but somewhat painful cut in the tongue by burning it with a cigarette . . ."

Sigmund Freud made his contribution in 1917. Listening to distraught Frau Emmy von N. discuss childhood traumas like being pelted with dead animals by her siblings, the eminent neurologist found the *frau*'s stutter to have been caused by "displacement upward of conflicts over excremental functions."

This insight was amplified in 1928 by Dr. Isador Coriat, an influential Boston neurologist and psychiatrist. Dr. Coriat discovered that stuttering was caused by a patient's childish desire to nurse:

> [T]he stammerer will be seen in the act of nursing at an illusory nipple . . . The stammerer will often bite the

> tongue . . . *a symptomatic cannibalistic feature that is a*
> *remnant of the early and primitive oral sadism . . . pro-*
> *longed oral possession tends to annihilate the word*
> *through . . . sucking and biting.*

Dr. Coriat also noticed that stuttering patients would sometimes gesture with their hands, something he found "peculiar." From this he was able to discern that stutterers stuttered because they were desperate for attention and wanted to be fed, or even wiped. He also noted that when mumbling certain consonants, stutterers used the same muscles as babies sucking milk from their mothers' breasts.

According to Coriat, stutterers "behaved" this way out of a childish need to gratify an out-of-control oral libido. And because most of Coriat's male stutterers happened to have both an Oedipal complex and a mother fixation, they were latent homosexuals as well. Dr. Coriat's female stutterers weren't much better. According to him, because of their castration complex, they tried to turn their tongue into a penis so they could play with it. Disgusted with themselves, they then came to hate their mother, the original castrator.*

---

* Similar notions continue to captivate psychoanalysts and plague stuttering patients. In 1999, the *International Journal of Psychoanalysis* published "Speaking in the Claustrum: The Psychodynamics of Stuttering":

> [T]he stutterer is working out intolerable experiences of separation from the primary object and a resulting catastrophic experience of the oedipal situation through an unconscious fantasy in which anal qualities are conferred on the internal maternal object by a predominating hatred. The intrusive identification of parts of the self in the maternal rectum gives rise to a claustrophobic experiential world . . . The anal-sadistic object space of the claustrum is projected on to the external object space and thus also on to the mouth as the origin of the sound envelope.

Stroke of Luck

For a lucky few, getting struck by lightning was just the cure for whatever ailed them.

Leconte on the Effects of Lightning, published in the *New York Journal of Medicine*, wrote of an 1843 incident in which, even taking into account "the vagueness of [her] negro testimony," it was certain that a seventy-year-old woman had been struck by lightning near a mulberry tree. The woman was found revitalized, her aging process reversed. Medically speaking, her newfound vigor was attributed to the lightning's effect on the ganglionic nerves of the solar plexus.

In the same journal, another doctor wrote of a deaf boy. After being struck by lightning in the ear, he had a cup of tea and could hear. The *American Journal of Science and Arts* reported the story of Samuel Leggers, paralyzed in the face and virtually blind. A day after he was struck by lightning, he regained his vision and wrote long letters. In 1823, the book *Cure of Asthma by a Stroke of Lightning* was published.

The *Lancet* published "Therapeutic Effects of Lightning Upon Cancer" in 1880. It described how a farmer with cancer was struck by a bolt of lightning that entered his trousers, killed his horses, and splintered his plow. Knocked unconscious, the lucky farmer woke up to find his cancer miraculously in remission. According to the article, it was only a matter of time before "frictional electricity" would become "one of the most powerful therapeutic agents in the dispersion of cancerous formations."

Our Presidents

Like King Charles II of England, American presidents received their own royal treatment.

On December 13, 1799, George Washington complained of a bad sore throat, and the next morning he had trouble breathing. He forced a reluctant servant to take a pint of his blood, saying, "Don't be afraid . . . More, more." Three doctors then arrived, among the most respected in the nation. The first blistered Washington's skin with dried beetles and did two bloodlettings of twenty ounces apiece. Fearing that wasn't enough, he took forty ounces more.

Another doctor arrived; he took thirty-two ounces. In all, over a period of ten hours, about four quarts were taken, more than half of Washington's blood volume. At 10:10 p.m., Washington died.

Our ninth president, William Henry Harrison, caught a cold that developed into pneumonia. He was blistered and cupped (suctioned of blood), forced to vomit and purge, and given opium, brandy, and Virginia snakeweed. Our twelfth president, Zachary Taylor, ate too many berries. He got the same treatment, and he died too.

Abraham Lincoln suffered from "melancholy." Doctors and historians believe Lincoln was taking the "blue pill," prescribed by nineteenth-century physicians for just about everything, including melancholy. The pill contained mercury, a powerful neurotoxin. Taken the normal two or three times a day, it would have delivered a dose nearly ten thousand times today's accepted level.

The pills likely exacerbated a preexisting tendency toward melancholy. Henry Clay Whitney, a young lawyer just learning his way around the courthouse, saw Lincoln brooding by himself, in a corner.

> [N]o relief came from the dark and depressing melancholy, till he was roused by the breaking of the court, when he emerged from his cave of gloom and came back, like one awakened from sleep.

After Lincoln's assassination, the fierce chief of a remote Cossack tribe asked Russian novelist Leo Tolstoy to tell the great man's story. On being handed a photograph of Lincoln, the warrior said, "Don't you find, judging from this picture, that his eyes are full of tears and that his lips are sad with a secret sorrow?"

Lincoln also had a volcanic temper. Of the fourth of the famous Lincoln-Douglas debates, Ward Hill Lamon, Lincoln's bodyguard, wrote:

> [His] whole frame shook ... Mr. Lincoln reached back and took Ficklin by the coat-collar, back of his neck, and in no gentle manner lifted him from his seat as if he had been a kitten, and said: "Fellow-citizens, here is Ficklin, who was at that time in Congress with me, and he knows it is a lie." He shook Ficklin until his teeth chattered. Fearing that he would shake Ficklin's head off, I grasped Mr. Lincoln's hand and broke his grip.

Ficklin was a close friend. Afterward, he said, "Lincoln, you nearly shook all the Democracy out of me to-day."

Just before Lincoln assumed the presidency, he supposedly stopped taking the pill. Acclaimed for his steady hand and "infinite patience," he calmly steered the country through its greatest crisis.

Surgical Theater

Public dissection was fine, but by the nineteenth century people began to prefer the live theater of actual surgery, done on patients who, before the age of anesthesia, were fully awake. Doctors, "professional gentlemen," and anyone with a few minutes to spare were encouraged to stop by. Anything could happen.

In 1831, with some fanfare, happy-go-lucky Chinese laborer Hoo Loo made his arrival in London. Loo came to see Dr. Aston Key, one of the best surgeons of his time. Loo had a tumor overhanging his groin, which, at sixty-five pounds and four feet in circumference, was one of the largest ever seen.

Loo's surgery was supposed to be done in the operating room, as usual. But the mob outside, flashing "Hospital Tickets," grew so unruly that the procedure was switched to the Great Anatomy Theater, used for dissections. It held two hundred people. When the surgery began, five hundred to seven hundred "gentlemen," in top hats and tails, jostled for position.

After eating a large breakfast, an affable, joking Hoo Loo, accompanied by his interpreter, walked into the theater. He was laid on the dissection table, blackened with blood from previous, less fortunate occupants. A handkerchief was placed over Loo's face and his hands were tightly bound. Key worked the crowd, describing the nature of the tumor and the surgery he was about to do. He received a polite round of applause.

Nurses pumped brandy into Loo's stomach to keep him awake, and the procedure began in earnest. Loo's penis was cut off, but, in all, his screams were said to be "no more than expected." At least once, perhaps several times, according to varying accounts, the crowd burst into applause. For those who couldn't get in, bulletins were issued every half hour just outside the hospital.

Things went well. Then an artery began spurting. Loo yelled out, "Unloose me! Water! Help! Water!" An athletic-looking man in the audience was chosen to donate blood, and Key got the bleeder under control. The tumor was finally cut out, but Loo died anyway. Newspapers reported that he was a victim of heat dehydration, supposedly caused by the overflow crowd. Loo's last words were, "I can bear no more!"

Ashes on the Ground

So she was burnt with all her clothes,
And arms and hands and eyes and nose;
Till she had little more to lose
Except her little scarlet shoes
And nothing else but these were found
Among her ashes on the ground.

—*German psychiatrist Heinrich Hoffman (nineteenth century)*

Spontaneous human combustion was a tragedy. One man had flames shoot from his nostrils. Thomas Williams burst into a flame of bright blue, and fire leaped from the stomachs of two noblemen drinking at a stag party. On several occasions, after eating too much, a "gross feeder" awoke to find himself engulfed in a luminous halo.

Nineteenth-century medical journals reported these incidents and many others. One concluded that spontaneous human combustion was a slow, gradual process, based on the man who lay atop a giant oven, only to wake up and find that his arm burned off. Most comprehensive, perhaps, was the study of twenty-eight human ignitions published by the *Boston Medical Journal*. It concluded that victims tended to be very fat, providing more fuel for the fire, or very skinny, making them dry, like tinder.

What actually caused the phenomenon was anybody's guess, and experiments proved unreliable. A study of fireflies went nowhere. Watkins tried to set fire to the body of a pirate, but it didn't burn after several hours, and he went home.

Theories abounded. In the *Lancet*, Jöns Jacob Berzelius favored neutralization of opposed electricities, but F. J. A. Strubel believed flames were triggered by the breakdown of water molecules into flammable oxy-hydrogen gas. Dr. B. Frank of Göttingen thought combustion was caused by subtle thermodynamic reactions among the body's phosphoric compounds. Based upon a Royal Society of Paris report of a butcher who got burned when he cut the bloated body of an ox, some thought combustion was caused by the ignition of hydrogen sulfide within the body cavity.

By the twentieth century, in respected medical journals at least, the phenomenon had burned itself out.

Childbed Fever

The great Ignaz Semmelweis died alone and annonymous, as was intended.

When Semmelweis became chief resident of Vienna General Hospital in 1846, women giving birth there dropped dead so fast and so often that more savvy mothers chose to deliver at home.

In the hospital, when a dead mother's body was carted from maternity to morgue, the corpse would often be found full of "clotted milk." The "milk," which we now know was infected pus, was thought to be caused by an "aura," or "atmospheric influences," hanging like a cloud over the hospital.

Semmelweis didn't know about germs, but he did know that the same doctors who cut open dead mothers in the morning delivered live babies in the afternoon, without washing in between. He suspected it was something from these doctors, not some mysterious vapor, that was killing the patients.

Semmelweis insisted that the doctors wash themselves, but they were insulted and refused. Among older, more established medical men, wearing a crusty, bloodstained smock and carrying that "hospital odor" was considered a badge of honor.

More open-minded interns began washing their hands, and mothers stopped dying. But the hospital's more experienced doctors, and the *Viennese Medical Journal,* ran Semmelweis out of town. More mothers died.

In 1865, Semmelweis was put in a straitjacket and locked away. When he died, apparently from an infection he got on the ward, his native Hungarian Association of Physicians refused to print his obituary. His sad, sordid end proved, once and for all, that he had been crazy all along.

Semmelweis's house is now a museum. An Austrian coin bears his likeness, and Viennese women give birth in safety at the Semmelweis Klinik.

The Body Electric

Powerful and edgy, electrical current was the perfect tool for amped-up doctors of medicine's Heroic Age. In 1820, one London clinic reported four thousand patients "cured" by electricity. Results were so encouraging that in 1836, Guy's Hospital built an "electrifying room." It was mostly for poor people, who could be experimented on.

A turn of the dial was just the thing for rundown bodies, especially given the discovery of electricity in the nervous system. As a Chicago doctor wrote, with the brain a positive pole and the nerves a negative, humans were little more than a large "voltaic pile."

As an electrical device, the human body was always being interfered with by static from other electrical devices. In an 1879 talk before the Baltimore Medical Society, Dr. George Beard, the foremost "electrotherapist" of his time, attributed an increase in nervous disease among "brain workers" to Thomas Edison and his newly invented electric light. More wattage, he believed, led to more mental disease.

Beard eventually became good friends with Edison, who himself began inventing things like the inductorium, which cured rheumatism with electricity.

By the 1880s, "the Golden Age of Electrotherapy," electricity was applied to just about any body part that hurt and was used to treat conditions such as bad eyesight as well. Wall-mounted power devices were standard equipment in doctors' offices, and patients stripped down for sessions in the "galvanic bath" and the "magnetic machine."

Just a jolt or two could cure impotence. Some doctors fitted a zinc-lined cylinder to a patient's penis and zapped away, but this procedure was improved upon by the great Beard himself, who made the device self-activating. As per the multivolume *A System of Electrotherapeutics* (1902), edited by S. H. Monell, "Professor of Static Electricity" and secretary to the American Medical Association:

> [T]he faradic brush applied to the testicles ... gives excellent results ... twenty or thirty applications are require to effect a cure ... 1 or 2 minutes on each side; the current should be ... strong, so that a distinct burning sensation is produced.

Sometimes the anode would be inserted up the tip of the penis.

A hundred years earlier, a scientist named Benjamin Franklin flew his kite during a thunderstorm and become renowned for his electrical expertise. When sick people requested that he harness this great power, Franklin, known for his common sense, turned them away.

With Pleasure

> When these symptoms indicate, we think it necessary to
> ask a midwife to assist, so that she can massage the gen-
> italia with one finger inside, using oil of lilies... And
> in this way the afflicted women can be aroused to the par-
> oxysm.
>
> —*OBSERVATIONEUM ET CURATIONEM MEDICINALIUM*
> *AC CHIRURGICARUM OPERA OMNIA* (1653)

From at least the time of the Greeks and Romans, doctors
treated "hysteria" using manual stimulation. Inducement
of a "spasm" was thought to alleviate the illness, at least
until the next session. The treatment was strongly recommended for nuns and widows.

Nineteenth-century doctors found the task unpleasant
and didn't know how to do it anyway, so they would generally hand it off to midwives and assistants. Eventually, en-
terprising physicians invented mechanical devices to
do the job for them.

In the 1860s, an American doctor designed a steam-
powered pulsating device with a vibrating sphere. In 1890,
a British physician patented a superior electromechanical
device. Hydrotherapy also came into vogue, with medical
spas installing high-pressure showers and faucets.

Doctors employing these devices had to proceed with
caution. An American doctor warned that use of his own
"pelvic manipulator" was to be closely supervised to prevent overindulgence. In 1843, French physician Dr. Henri
Scoutetton warned:

The first impression produced by the jet of water is painful, but soon the effect ... create[s] for many persons so agreeable a sensation that it is not to go beyond the prescribed time.

Manual and mechanical stimulation for hysteria remained standard medical practice until the 1920s.

Sulpheretted Hydrogen
(to Be Polite)

Having already written of the patient who smoked a cigar and almost blew himself up, Sir Lauder Brunton was well aware of the dangers of "sewer gas" lurking deep inside us.

Knighted for his pioneering work with the circulatory system, Brunton turned his attention to the health crisis seizing Britain's intellectual upper crust, its "brain workers." In contrast to hardy physical specimens of yore, Englishmen of good breeding now required constant rest and attention, and sometimes even a long cruise on a luxury steamship. Brunton wrote of good men gone bad, and hard men gone soft, and sober men turned to tippling.

Bringing his considerable medical expertise to bear, Brunton discovered the cause of the ruinous epidemic—sulpheretted hydrogen. Smelling like rotten eggs, the gas ruined the minds and broke the spirits of England's best and brightest.

In a series of well-received articles and lectures, Brunton detailed how fumes of "poisonous peptones," a powerful "protoplasmic poison," built up in the intestine and congealed in the nerve centers. Left untreated, the fumes would pollute the brain and cause depression, hysteria, and assorted mood disorders.

Worst of all, unlike fumes from a person's behind, gas on the brain had nowhere to go, because the brain was encased in a rigid skull. As corroboration for his theory, Brunton cited experimental work done by both Albertoni of Genoa and Schmidt-Mühlheim of Leipzig, under the auspices of well-known Professor Ludwig.

Lecturing before a crowd of rapt doctors in 1885, Brunton, with an air of resignation, noted that the problem was long-standing, having existed since cavemen first roasted the marrow of the cave bear and savored the meat of the woolly rhinoceros. There were, he said, no easy answers.

The Doctor's Riot

A poor little boy, shattered, walked past the hospital where his mother had just died. A sneering medical resident waved his mother's severed arm out a window and screamed, "Get lost, or I'll smack you over the head with it."

This is just one of many stories describing what triggered New York City's "Doctor's Riot" in 1788. But judging from a letter sent by Colonel William Heth to Virginia governor Edmund Randolph, the boy's version probably wasn't too far from the truth.

> *The Young students of Physic [medicine] have for some time past, been loudly complained of, for their very frequent and wanton trespasses in the burial grounds of this City. The Corpse of a Young gentleman . . . was lately taken up . . . A very hand-some & much esteemd young lady, of good connections was also, recently carryd off.*
>
> *On Sunday last, as some people were strolling by the hospital, they discovered a something hanging up at one of windows . . . part of a man's arm or leg tumbled out upon them. The cry of barbarity was soon spread—the young sons of Galen fled in every direction . . . one took refuge in the chimney . . .*
>
> *In the Anatomy room, were found three fresh bodies—one, boiling in a kettle, and two others . . . with certain parts of the two sex's hanging up in a most brutal position. These circumstances . . . exasperated the Mob beyond all bounds—to the total destruction of every anatomy in the hospital.*

The army was called out but retreated when the mob "[b]roke their guns to pieces, and made them scamper to save their lives." Ultimately, more than sixty guns were discharged, and several of the rioters were killed.

Dr. Clutterbuck's Lectures on Bleeding

In 1838, the first of "Dr. Clutterbuck's Lectures on Blood-letting," a five-part series, was published in the *London Medical Gazette*. Referring to the bloodletting of children, Dr. Clutterbuck wrote:

> *You should do too much, rather than too little ... By acting with such promptitude and decision you must expect occasionally to encounter opposition ... You may be accused of acting with unnecessary rigor ... do it with confidence, and without timidity.*

Dr. Clutterbuck cited an adult patient, Mrs. M., to whom he successfully applied over a thousand leeches, and a "generally healthy man" from whom, as a tonic, he often bled a pint a day.

What Dr. Clutterbuck was writing, everyone else already knew. In 1824, a French soldier fainted from blood loss after being stabbed in the chest. He was immediately bled of twenty ounces more, "to prevent inflammation." Later that night he was bled of another twenty-four ounces, and the next morning ten ounces more. In the next fourteen hours, he was bled five more times.

There was more bloodletting over the next few days, and his doctors applied first thirty-two and then forty leeches. Somehow, the soldier survived both his wounds and the bleedings. His supervising doctor wrote, "[B]y the large quantity of blood lost ... the life of the patient was preserved."

The Stethoscope

As before, nineteenth-century medicine was a profession of learned gentlemen. Beyond checking a pulse or taking a temperature, doctors of discernment looked upon a patient's body with something approaching disdain. Dissecting a corpse as a student was a necessary evil, but when dealing with a live patient, a respectable doctor knew how to keep his distance, and with it, his social standing. Bodywork was better left to the boorish surgeons and their ilk. The sounds of a sick body were likewise to be ignored and were considered an embarrassment to both doctor and patient, especially among the upper crust.

In 1816, this doctor-patient gap was bridged, tastefully, by René Laennec of France. Examining an obese woman with a heart problem, he wanted to actually listen to her heartbeat but was too embarrassed to touch his ear to her abdomen. He rolled sheets of paper into a cylinder and touched one end to her heaving chest. The other he put to his ear and listened. With the first-ever stethoscope, he magnified the sound of her heartbeat and lungs. Of equal importance, he preserved both his dignity and her privacy.

Doctors resisted Dr. Laennec's intrusive invention for decades. A lecture published in the *London Medical Gazette* argued that the stethoscope was no better than the unaided ear. In 1848, Harvard medical professor Oliver Wendell Holmes, father of America's greatest Supreme Court justice, composed a ditty, "The Stethoscope Song." It compares the instrument to a baby's rattle and describes the plight of an earnest young doctor who fatally misdiagnoses patients when a spider and two flies fall into his shiny new stethoscope.

> *Now use your ears, all that you can,*
> *But don't forget to mind your eyes,*
> *Or you may be cheated, like this young man,*
> *By a couple of silly, abnormal flies.*

With reviews like this, most English and American doctors took a pass. Others wore the stethoscope with pride but didn't know how to use it. One doctor impressed his patient with a diagnosis of respiratory murmur but applied the wrong end to his ear and reported the sound of a passing horse and buggy.

Even the instrument's admirers thought it a bit strange. Later models, the kind we're familiar with, dangled loosely from a doctor's ears. They were thought to look so silly that doctors became self-conscious, and a few more sensitive souls took to wearing them coiled under their hats. When the stethoscope of a medical student fell to the ground during a snowball fight, he was accused of weapons possession.

Patients were also confused. Some thought the stethoscope a new type of surgical tool and feared they were about to be opened up. Others, a bit more clearheaded, saw the instrument's savage potential:

> *Stethoscope: thou simple tube*
> *Clarion of the yawning tomb*

Of course, privacy continued to be an issue. One stethoscope featured a tube several feet long so the doctor could listen from another room.

Eventually, patients shunned doctors who didn't use the stethoscope. But given the state of medicine at the time, even using the instrument, there often wasn't much a doctor could do anyway.

In 1826, at age forty-five, René Laennec died of tuberculosis and heart complications. With the stethoscope, his doctors would have been able to monitor the progress of his disease, and slowly watch him die.

Awful Effects

Only in the eighteenth century did "self-pollution" become an actual medical problem, held responsible for everything from epilepsy to drooping shoulders. Nineteenth-century publications such as "Facts and Important Information from Distinguished Physicians and Other Sources: Showing the Awful Effects of Masturbation on Young Men" revealed that "neither plague, nor war, nor small-pox, nor similar diseases, have produced results so disastrous to humanity as this pernicious habit."

The Influence of Sexual Irritation Upon the Diseases of the Ear (1884), translated by the Louisiana Medical Society from the original German, told the story of a naïve country boy. Attending college for the first time, he suffered a perforated eardrum. The boy's symptoms, his doctor wrote, "caused me to suspect that he had acquired the habit of masturbation." He had been initiated by a "vicious roommate."

Boys and men were the usual culprits, but females did not go untouched. In 1867, the *British Medical Journal* published "Influence of the Sewing Machine on Female Health." At London Hospital, Dr. J. Langdon Down noticed that seamstresses exhibited "diminished luster" and complained of cobwebs before their eyes.

> [T]he symptoms met with among machinists was not due to machine labor per se but to immoral habits, which had been induced by . . . the movement of the legs . . .
>
> [I]f machines are employed, these should be selected where the motor power in a manner not liable to produce local hyperaemia [increased blood flow].

The consensus was universal and the symptoms easily observable. The superintendent of a well-known mental hospital wrote:

> In this yard were ten or fifteen persons, most of whom had become insane by this indulgence. One of these, a man of bloodless countenance and vacant gaze, was promenading back and forth as fast as his feeble limbs could bear him. O, how fallen from the talented and accomplished man he once was!

Masturbation was a mechanical problem begging for a mechanical solution. In the early 1850s, Boston doctors devised a spike-lined penis shield, to be attached to the body with a steel band. A thirteen-year-old boy tore through his band, but the rivets held, and Dr. Fleck refitted him with a new one. The Stephenson Spermatic Truss, patented in 1876, made sure the penis was always pointed down, while the Bowen Device, attached with chains and clips to pubic hair, insured pain on erection. The 1908 *De-*

troit Medical Journal advertised a "Timely Warning Ring," made of lightweight aluminum, not brass, adjustable to any size organ, and featuring a wide opening hinge.

The 1885 *Handbook of Medicine* described a metal cage that allowed an erection but prevented touching; some similar devices had padlocks on them. Doctors also deployed "Sexual Armor," invented by sanitarium nurse Miss Perkins. Her breakthrough featured a leather jacket supporting a large piece of steel armor with a hole for urine to escape. To defecate, the wearer needed a helper to unbolt the device from the back.

More drastic solutions were proposed. In an 1883 edition of the *Boston Medical and Surgical Journal*, Dr. Timothy Hays cited three cases in which he had surgically resected the spermatic ducts: "The sexual appetite was as effectively destroyed as by castration." Dr. John Harvey Kellogg, best known for his invention of the cornflake, wrote:

> [Circumcision] should be performed . . . without using an anesthetic, as the brief pain attending the operation will have a salutary effect upon the mind . . . In females, the author has found the application of pure carbolic acid to the clitoris an excellent means of allaying abnormal excitement.

In 1876, Dr. Abraham Jacobi, later to become president of the American Medical Association, advocated infibulation (surgical mutilation) of the penis and the infliction of "artificial sores." In 1902, Dr. Wintertauk took a more gentlemanly tack. He suggested "the little sinners" be tied spread-eagled across the bed or made to wear mittens. Across the ocean, led by Claude François Lallemand, the noted French surgeon and medical professor, the *Lancet* published more than thirty articles discussing recent findings in the fields of masturbation and spermatorrhea (wet dreams).

Of a young man suspected of shameful habits, one doctor wrote:

> [T]he apparent languor, the downcast, unquiet look, and hypochondriac expression of the patient, and my suspicion was at once awakened . . . I requested his mother, who accompanied him, leave the room . . .
>
> [U]pon proceeding to examine the urethra, the same cowardly dread of pain which is common in people with those habits was strongly shown.
>
> A bougie [metal cylinder] was passed up the urethra . . . he regularly screamed out.

After having a "caustic" applied to his groin, the young man never returned. "[A]ll has gone well," the doctor reported.

Lallemand himself liked to stick a catheter up the bladder of wayward boys to create swelling and blockage. As with the other doctor, all was well. "[P]atients invariably experience a sense of relief immediately after its removal," he wrote.

What God Intended

[Y]ou must expect to suffer . . . you will suffer—you will suffer very much.

—DR. M. DUBOIS, TO PATIENT AND FRIEND FANNY BURNEY

In 1811, before the era of anesthesia, prominent author Fanny Burney underwent breast cancer surgery. As planned, surgeon Dubois and a SWAT team of doctors prepared in secret, and, to lessen the dread, contacted her just two hours before.

[T]he clock struck three . . . [My room] was entered by seven men in black . . . the maid, however, and one of the nurses ran off . . .

A terror that surpasses all description . . . the dreadful steel was plunged into my breast, cutting through skin, vein, flesh and nerves . . . I began a scream which lasted unremittingly through the whole operation . . . I felt the instrument, describing a curve, cutting against the grain . . .

[I] thought the operation was over—but no! . . . the terrible cutting was renewed—and worse than ever . . . yet all was not over . . . I felt the knife racking against the breast bone, scraping it . . . again started the scraping, attom after attom . . .

When I opened my eyes I saw the good Dr. Larrey, pale nearly as myself, his face streaked with blood, and its expression depicting grief, apprehension and horror.

Through the Renaissance, the lessening of pain, especially during surgery, was looked upon with suspicion. Surgical procedures were often fatal, and with God and the devil fighting it out for the soul, a conscientious doctor made sure his patient was in full possession of his wits.

In the sixteenth century, Grand François was tortured on the rack. It was noted with some disapproval that after eating soap to anesthetize himself, "[h]e fell asleep, and the toes were torn from both his feet without his manifesting any signs of pain . . ." Around the same time, Lady Macalyne of Scotland asked a midwife to ease her pain while delivering twins. For this, the lady was burned alive.

In the nineteenth century, the vice president of the American Medical Association declared, "Pain is curative," and some doctors wouldn't operate until their patients came out of shock or coma. Outlawing anesthesia, the city fathers of Zürich proclaimed, "Pain is a natural and intended curse of the primal sin. Any attempt to do away with it must be wrong."

An 1849 article in the *Lancet* declared that the painless delivery of babies was an invention of the devil. The president of the American Dental Association wrote:

> [A]nesthesia is of the devil and I cannot give my sanction to any Satanic influence . . . I do not think man should be prevented through what God intended them to endure.

Mad Dogs

The Talmud says a person bitten by a mad dog may not be given the lobe of its liver to eat. This passage was once contested by a prominent rabbi who, like Galen, believed in liver's power to heal. However, the rabbi was forced to concede that eating a dog would not be kosher. The Talmud has much else to teach us about rabies. It tells us that dog bite victims should write upon the skin of a male hyena, strip naked, bury their clothes for a year, burn the clothes, and scatter the ashes.

Real medicine didn't do much better, as illustrated by hundreds of articles, letters, and case histories published by the *Lancet*. For the bite itself, doctors recommended pouring boiling oil on top, or applying a red-hot iron, or igniting gunpowder in the wound. Some doctors noticed that drowned dogs seemed to be rabies-free. They half drowned their patients in the ocean and hired boatmen with poles to fish the semiconscious bodies out of the water. In Greece, doctors had patients bathe in water used to clean gun barrels or in the juice of a crawfish. Dr. Marshall Hall observed that frogs injected with poison recovered when left in peace, so he simply laid his rabies patients down on comfortable beds. This became standard practice for decades.

Several writers attributed rabies to inadequate sexual release, and at least one physician suggested that doctors castrate their patients. The *Lancet* also recommended seating patients in a wicker chair over hot bricks, dancing, and massive doses of asparagus.

Proceedings

Pursuant to minutes taken in 1886, present at the first meeting of the Association of Medical Officers of the American Institutions for Idiotic and Feeble-Minded Persons were doctors Sequin and Wilbur of New York, Doren of Ohio, Wilbur of Illinois, Knight of Connecticut, and

Kerlin of Pennsylvania. Dr. Wilbur of New York was appointed chairman and presented "The Agenda on Education and Training of Idiots and Imbeciles." The following is in paraphrase:

JUNE 1, 1876

At Keystone Hall, the Association was entertained with music, marching, and calisthenics by the children and ladies of the Training School, with Drs. Parish and Tuck of the Massachusetts School for Idiots also in attendance. Officers were elected, and a Constitution was adopted.

JUNE 8, 1876

Over objection by Drs. Wilbur and Kerlin, a resolution allowing ladies to become members was approved. Drs. Black and Jarvis, who were unable to attend, expressed their regrets.

JUNE 12, 1877

At the Ohio State Asylum for the Education of Idiotic and Imbecile Youth, the meeting commenced at nine p.m., with many guests and inmates in attendance. Dr. Wilbur presented his paper, "Classification of Idiots."

JUNE 13, 1877

All were present. Typical cases of idiocy were discussed, including those of the Mongolian, Cretin, and Aztec types. Dr. Brown proposed that because of reversion to higher types, drunkenness was dying out.

JUNE 14, 1877

Congratulatory letters were read. Dr. Wilbur proposed that laws be passed providing insane idiots the same admission privileges to insane hospitals that other insane persons have.

In the afternoon, the association visited the Central Hospital for the Insane and the Deaf and Dumb Asylum. In the evening, the association, at the invitation of Professor Snead, was entertained by pupils of the Blind Asylum, who had prepared a most excellent opera for the occasion.

JUNE 15, 1877

Sections A, B, and C of Dr. Kerlin's paper, concerning the idiotic and imbecile classes, was deferred for consideration until June 1878. The treasurer reported a balance of sixty dollars on hand. On motion, the association proceeded to elect officers for the next year.

Mrs. C. W. Brown presented a paper, "Prevention of Mental Disease," noting that recently an otherwise sensible businessman had been overheard speaking proudly of his idiot child. Mrs. Brown proposed that higher civilization unwittingly aided the propagation of the unfittest, and that soon children possessing robust bodies, keen intellects, and pure souls would be replaced by witless unfortunates occupying crippled bodies and having dwarfed souls.

A Sudden Burst of Insanity
Among the Colored Race

We still memorize taxonomy in high school—phylum, genus, species. The classification system was created by Carl Linnaeus, a Swedish medical doctor.

Dr. Linnaeus divided the human race into four main racial branches—*Americanus* (red-skinned, stubborn, easily angered); *Asiaticus* (avaricious, diabolical); *Africanus* (careless, unreliable); and *Europeanus* (clever, stable, really very nice). Subraces were designated wild men, dwarfs, troglodytes (primitives), and "lazy Patagonians" (South Americans).

In the 1851 *New Orleans Medical and Surgical Journal*, Dr. Samuel Cartwright, honored for his work with cholera, announced his discovery of a new disease, *drapetomania*. Among the symptoms were "[the] troublesome practice that many negroes have of running away." A related disease, *dysaesthesia aethiopica* (rascality), was held attributable to excessive liberty.

Many Southern doctors believed that African Americans felt little if any pain. In an article in the 1854 *Virginia Medical and Surgical Journal*, a doctor wrote that, to cure a slave of pneumonia, he threw five gallons of boiling hot water on his spine. The doctor was surprised that the treatment "rouse[d] his sensibilities somewhat, as shown by an effort to cry out." In an article concerning "Distinctive Peculiarities," the *Memphis Medical Recorder* claimed that African Americans, while good at hearing, seeing, and smelling, couldn't feel very much. "[T]hey submit to the rod with a surprising degree of resignation and even cheerfulness," it reported.

Dr. J. Marion Sims, the father of American gynecology,

bought or rented at least ten slave women to perfect his procedure for correcting vaginal fistulas. Three underwent multiple surgeries, without anesthesia. One, named Anarcha, had thirty. Sims wrote that the women had been "clamorous" for the operations, and, fully awake, even helped him while they were being operated on. A statue honoring Dr. Sims now stands in New York's Central Park.

Southern physicians examined thousands of pages of Census Bureau data. Painstaking research revealed that freedom was bad for a black person's health. In fact, the abolition of slavery had caused "a sudden burst of insanity among the colored race." The numbers were irrefutable. Dr. T. O. Powell of the Georgia Lunatic Asylum concluded:

> *The census of 1860 [pre–Civil War] will show . . . one insane negro in every 10,584 . . . The census of 1870 [after] shows . . . one to every 4,225 of the population.*

After much medical debate, the claim, based on bogus numbers and bad science, was exposed as a fraud, most notably by James McCune Smith, the first African American doctor granted a medical degree, and, coincidentally, a trained statistician.

The Mouths of Babes

In 1839, teething took the lives of 5,016 of London's babies, according to the city's registrar general.

Dr. Jacob Plank, a founder of modern dermatology,

believed that teething caused lameness. London's leading surgeon, John Hunter, thought teething responsible for discharge from the penis.

To stop the carnage, doctors went after baby teeth and gums with everything they had. In 1844, the *Lancet* told doctors to "scarify" with a sharp knife every single baby tooth once, if not twice, a day. Another journal bemoaned the "superficial incisions" of the timid and recommended cutting teeth and gums right down to the bone.

Swedish physician Rosen von Rosenstein employed leeches, which gnawed with their own little teeth. But the gum lancet, carried in a doctor's back pocket, was generally the weapon of choice. As a prophylactic, some babies got a burning cautery iron to the back of the head.

To stop all the crying, doctors recommended patent medicines like Woodward's Gripe Water, made of alcohol, and "The Mother's Friend," Mrs. Winslow's Soothing Syrup, which contained morphine and sometimes created baby addicts. In the United States, it was held responsible for a rash of infant comas and deaths. Butler's Electro-Medical Teething Necklace may have been scary, but it was medicines containing mercury that caused an epidemic of "pink disease" (mercury intoxication). This killed between 10 and 25 percent of the babies it afflicted.

Doctors, not teething, were responsible for the London deaths. Most of the five thousand babies died from infections caused by dirty instruments.

Pound of Flesh

When an African tribe flayed Jonas Wright alive in 1632, they bound a book with his skin and presented it to his friend as a keepsake. This was anthropodermic bibliopegy, a practice later adopted by many doctors.

It wasn't confined to medical men; skin from victims of the French Revolution was used to bind books propounding the rights of all men. More congenially, when John Fenno courageously resisted James Allen's attempt to rob him on the Massachusetts Turnpike, Allen was so impressed that he bequeathed a book bound with his own skin to Fenno, "as a token of his esteem."

By the nineteenth century, anthropodermic bibliopegy among doctors had become a quaint but respectable hobby. To illustrate the first diagnosed case of trichinosis, John Stockton Hough bound three medical volumes with his own patient's skin. Admiring colleagues found the cover "cheap, durable and waterproof." Naturally, a few dermatologists bound their books with skin.

Several texts ended up in medical schools and libraries. The John Hay library at Brown University houses Vesalius's classic *De Humanis Fabrica* (*On the Fabric of the Human Body*), bound in its own human fabric. London's Wellcome Library, the world's largest medical history collection, has several gynecological essays bound in human flesh.

Harvard's stacks, including the medical library, contain at least three volumes bound with human skin, and, pending DNA analysis, maybe many more. In interviews, Harvard librarians and archivists have made it clear that, for now at least, their collections are complete.

Sweet Dreams

Spermatorrhea (wet dreams) was masturbation's evil little twin. Men doomed to nightly recurrences flocked to doctors for treatment, hoping to dodge insanity, paralysis of the limbs, and even death. By the nineteenth century, demand was so great that some doctors thought relief would be available only to the rich and well connected.

Influential French medical professor Claude Lallemand, already known for his pioneering work in masturbation, dominated the field. Employing powerful microscopes to detect stray sperm, Lallemand concluded that spermatorrhea could be caused by hemorrhoids, bad weather, and worms. To curb the condition, he employed crushed ice, lead girdles, electrical shock, and hot needles to the groin.

> [T]he first of these needles is to pass through the raphe of the perineum . . . between the root of the scrotum and the margin of the anus . . . the third may be inserted in front of the first . . . I allow the needles to remain at least one hour, and at most three.

Lallemand also used a bougie, a slim metal cylinder with a ball on the end. Coated with a caustic substance and passed through the urethral canal, the instrument would deaden or destroy promiscuous nerve endings along the way. The procedure caused "visible agony" and the occasional death or two, but Lallemand thought it most effective. "[P]atients suffer such pain during the passage of the instrument," he wrote, "[it causes] the most marked and lasting effects."

Surgeon John Laws Milton was another pioneer. His classic textbook on spermatorrhea underwent eleven printings, the last in 1881.

Milton preferred Dr. Moriggia's hunger cure to Dr. Althaus's more complicated mechanical device, which ran electric current from the urethra's negative pole to the groin's positive pole. Milton himself injected skin-blackening silver nitrate into the groin and employed a four-pointed leather urethral ring, a device he used with astounding success on hundreds of patients:

> [I]t points turned inwards on the penis as to produce no uneasiness till erection comes on, when the patient ... can jump out of bed and thus arrest an impending emission.

Clever as they were, devices like this often worked their way off the penis in the middle of the night. For this, French physician M. Trusseau suggested using a padlock. Other doctors employed a mousetraplike "tooth ring," set to snap shut at the slightest provocation.

Dr. Trusseau also prescribed hot sand to the genitals and wool soaked in turpentine. Use of a metallic "stem pessary," he wrote, cured even the most "rebellious" of patients.

Unveiled at an 1871 meeting of the Medico-Physico Society of Florence was the *cintolo avissatore*, a type of leather sheath. Some physicians suggested adding a quick-release button, just in case. Another doctor in attendance recommended applying an electric sponge to the groin. To handle the mess, the patient would be seated in a cane-bottomed chair, with a slop pail placed underneath. In

1877, the *Bulletins et Mémoires de la Société de Paris* suggested putting spiked wooden strips on both sides of a bed. This would keep the patient from rolling over on his stomach, thought to be a leading cause of accidental emission.

Nothing could match the "electric alarum." "[A] ring placed on the penis is so made, that when expanded by erection it completes an electric circuit, and so rings a small alarum bell placed under the sleeper's pillow." More precisely:

> the instrument consists of a ring (A), which is hinged for the purpose of keeping the circuit open when the organ is quiescent. Upon the ring is a flat plate of ivory (a), furnished with a bolt (b), which, upon erection, is pushed backwards so as to complete the circuit. With the ring and bolt are connected two insulated wires (B, B) which convey the current two binding screws (C1, C2) . . . On the circuit being completed by the pushing back of the bolt (b), the hammer strikes the bell (G).

The device worked as intended, but after testing it at home several doctors complained. They said the alarm kept going off and that it was hard to clean.

The Brown Dog

Dr. François Magendie was a practical man. While colleagues debated whether the vital force animating all of life should be wedded to the theory of Newtonian gravitation, Magendie just wanted to see how things worked in the real world. Balzac wrote, "[Magendie] claimed the best medical system was to have none at all, and to stick to the facts."

Magendie was also a cruel man who enjoyed torturing animals. In 1825, Richard Martin, an Irish member of Britain's House of Commons, testified:

> [H]e got a lady's greyhound . . . nailed its front, and then its hind paws with the bluntest spikes that he could find . . . He then doubled up its long ears, and nailed them with similar spikes. He then made a slash down the middle of the face.
>
> After he had finished these operations, [he] then turned to the spectators and said, "I have now finished these operations on one side of this dog's head, and I shall reserve the other side till to-morrow. Although he may have lost the vivacity he has shown today, I shall [still] have the opportunity of cutting him up alive."

When he wasn't crucifying dogs, Magendie was poisoning them. To a lesser extent, he experimented on people too. Torturing animals gave Magendie a leg up in the race to distinguish sensory nerves from motor nerves. After a few horrifying live dissections, rival Sir Charles Bell had had enough. Magendie gleefully continued, using six-week-old puppies. So cruel was Magendie that even protégé Claude Bernard, the founder of modern physiology, agonized over the experiments. Bernard continued his research but did what he could for his subjects.

The battle for animal rights centered on a dog made of bronze. During the Brown Dog Affair of 1907, antivivisectionists in London erected a statue honoring a recently sacrificed pup. Crazed medical students, barking like dogs and howling like wolves, attacked the statue. Four hundred police quelled the riot. Both sides conducted riots around the dog for years, until, under cover of darkness, 120 police officers finally removed the dog.

In his later years, Magendie's interests thankfully shifted from torturing animals to botany. Tinkering in his garden, he lived to a ripe old age.

The Blushing Exam

Doctors of the nineteenth century shied away from touching their female patients or even looking too closely. An "examination" might consist of a woman, fully clothed, sitting in a comfortable chair and speaking about her illness, with a doctor "examining the expression of her countenance." Touching or even tapping the body was discouraged, though the taking of temperature and pulse was permitted.

No matter how dire the circumstance, "a blushing examination" was to be avoided at all costs. "Young doctors take such liberties," complained one properly prim lady to her doctor, Sir Arthur Conan Doyle. The speculum was invented by the Romans. When, finally, it came into widespread use, it was used on fully clothed women, while the doctor kept his eyes to the ceiling. In the middle of an exam in 1845, Dr. Marion Sims ran into a hardware store and bought a pewter spoon. Bending it into the type of speculum still in use today, he wrote, "I saw everything, as no man had seen before."

At a 1903 meeting of the American Medical Association, a Dr. Carsten said, "If young girls are left alone and live a moral life, they will not require the gynecological finger." It took one ailing woman eight days to admit to her doctor that she had a tumor. "She would not allow me to see it, but told me it was as big as a small hen's egg," her doctor wrote.

A doctor's reluctance to touch his patient wasn't always confined to women. Doctors were baffled as philosopher David Hume lay dying before their eyes, until a surgeon reached under Hume's shirt and discovered he had liver cancer. By the early 1800s, only five doctors in all of Paris were palpating (touching) their patients. In Padua, Italy, no physical examinations of patients were done until 1830.

"Treatment by correspondence," employed by many respected physicians, could avoid mortification on both sides. The patient, her privacy assured, would write a long letter describing her history and symptoms. Sitting in a comfortable leather chair in a book-lined study, safely shielded from his patient's body, her doctor would have the time and space to muse, cogitate, and reflect.

The American Journal
of Insanity

"Whether Deaf Mutes Are More Subject to Insanity Than the Blind" was only one of the hot-button issues addressed by the *American Journal of Insanity*, the go-to periodical for decades of mental health professionals.

More practical articles appealed to hospital superintendents and plant managers. Some, for example, had actual blueprints of plumbing systems capable of plunging patients from burning hot to freezing cold water in record time.

The workings of the mind received the most attention. One article weighed the pros and cons of swinging people from the ceiling to treat mania. "Spinning" was best avoided, it concluded, though it was "a very effectual means of producing sickness, vertigo and nausea." High-pressure "needle showers" would address "high cerebral excitement" and relieve the brain of its "surplus."

The ability to work up a good sweat, "cutaneous transpiration," was, according to the journal, the key to good mental health. Cited as proof were the tough-as-nails Romans, who sweated profusely in public baths. The journal also urged doctors to pump patients full of opium, malt liquor, strong beer, and arsenic and to apply "insanity blisters," corrosive bandages that created open, oozing wounds.

The journal was justly proud of its international bureau. It broke news that Italy's "Sirocco Wind" made Italians hot-blooded and that the *viente norte* (north wind) of Buenos Aires caused lazy Argentinians to take long siestas and get into knife fights. There, things had gotten so bad that women walked around with split beans on their foreheads to create blisters that kept them awake. "Insane foreigners" were overrunning our cities, it was reported, and a series of articles, "The Cretins of Switzerland," referred to those with Down syndrome. They were "frightful objects" ruining "an otherwise glorious country."

The *American Journal of Insanity* was eventually renamed the *American Journal of Psychiatry*, now the official journal of the American Psychiatric Association.

Ring of Fire, Redux

The medieval cautery iron was back and—among doctors, at least—more popular than ever. In *Archives of Medicine* (1879), Dr. E. C. Seguin wrote, "The ancient method . . . is still in vogue . . . burn deeply . . . use force in applying the instrument."

There was a problem—squeamish patients. When Dr. Brown-Sequard read his paper to the Massachusetts Medical Society in 1875, he said people were beginning to question "the suffering inflicted." A leading medical journal reported, "[P]opular dread of cautery is great, chiefly because of absurd accounts of burning patients . . ."

Doctors weren't unsympathetic. "I condemn the promiscuous burning of patients," one boldly declared. He suggested that doctors not cauterize patients by surprise, lest they be antagonized further. "I always tell my patients what I mean to do, how I mean to do it, and what the usual estimate of pain is."

A cautery iron made of platinum, which fit neatly into a large pocket, was the iron of choice for most doctors, but in 1876 an electric model was exhibited before the American Neurological Association. It was silent, not sizzling like the others, but was rather bulky, making it difficult to bring on house calls.

Some doctors plunged bone deep, as in the Middle Ages. Others thought superficial burns, one to six inches in length, were more beneficial. "[F]our to twelve strokes can be made in an incredibly short time . . . with very little suffering." Treatments every few days were considered best, and the cautery could be applied directly to the face with "perfect safety."

Dr. Brown-Sequard used the cautery to trigger epileptic attacks and revive patients from deep comas. He also used it for headache and sunstroke and for inflammation of the dura mater. The cautery also cured paralysis, he wrote, as long as the medulla oblongata wasn't too far gone.

Mice Pancakes

Pliny the Elder was shameless. For incontinence, he suggested eating a smaller fish found in a bigger fish's belly. But in *The Marrow of Physick*, Fourth Edition (1693), Dr. James Cook was so disgusted by his own bed-wetting cures that he wrote them in Latin so regular people couldn't read them.

Through most of medical history, doctors happily steered clear of incontinence, with folk medicine filling the vacuum. People ate roast pig penis sandwiches topped with buttered horse dung and stuck frogs to their kids' waists. Mice seem to have been a universal remedy; in disparate cultures all over the world, people had them fried, boiled, baked into pancakes, and worn around the neck.

During medicine's Heroic Era, doctors naturally took a more active role. They tied loops around boys' penises, for example. When medical journals warned of gangrene, they switched to metal clamps.

Later innovations included freezing genitals in ice, taking cocaine, and applying carbonic acid. Some pediatricians blistered the butts of children with noxious chemicals so they'd lie on their stomachs. When electricity came into vogue, electrodes were stuck up a child's rectum.

Sigmund Freud thought bed-wetting was caused by people's tortured psyches. Patient "Dora," for example, wet the bed as a substitute for genital gratification. Freud, a cigar smoker, also connected bed-wetting to fire. This explained why, he was certain, Dora always wanted to kiss him.

In later years, experts in the field steered a middle course, prescribing gentle therapy and absorbent sheets.

Big-Game Hunter

John Hunter was a grave robber as a medical student, and even after becoming Europe's top surgeon he continued his collecting. Dr. Hunter didn't much care where he got his body parts, or from whom. In 1791, famed composer Joseph Haydn saw him about a nasal polyp. Of the consultation, Haydn wrote:

> He inspected my polyps and offered to cure me of the disease ... Mr. Hunter asked me to visit him due to some urgent circumstances. I went there. After the opening compliments some robust fellows entered the room, seized me and tried to force me on a chair. I yelled, punched and trampled with my feet until I managed to free myself. Mr. Hunter was already in stand-by, with his surgical tools.

Haydn had designs on Hunter's wife, but Hunter had designs on Haydn's polyp, so he could put it in his museum of medical oddities.

Haydn's polyp was nothing compared to Hunter's campaign to obtain the body of seven-foot, eight-inch Charles O'Brien. O'Brien happened to be alive at the time, but Hunter bided his time, with a huge copper pot waiting to boil him down. Hunter demanded O'Brien's body upon his death, but the impertinent giant refused to give it to him.

Knowing that O'Brien was ailing, the wily Hunter, with contacts all over town, had him followed. Meanwhile O'Brien, aware of Hunter's scheming, hired an undertaker to ensure that when he died his body would be put under guard, sealed in a lead casket, and thrown into the sea.

The hapless giant never stood a chance. When he died, Hunter and his henchmen paid off the undertaker, bribed the guards, and snatched the body from the casket. The empty casket was then filled with rocks and dumped in the ocean.

The ruthless Hunter beat out other prominent London doctors, who wanted O'Brien for their own collections. The *Morning Herald* reported, "The whole tribe of surgeons put in a claim for the poor departed Irish giant, and surround his house, just as Greenland harpooners would an enormous whale."

Triumphant, Hunter boiled down his trophy in the giant pot. O'Brien's bones are now on display at the Hunterian Museum, affiliated with London's Royal College of Surgeons.

Our Faithful Friend

Overcome with emotion, in 1841 physician J. D. Snodgrass dedicated a lengthy poem to medicine's most revered instrument. "To My Spring Lancet," it was called. One line went, "I love thee, bloodstained, faithful friend."

Bloodletting reached its peak during medicine's Heroic Age. Doctors relied on it, and patients came to expect and even demand it. Holding a patient's arm with grace and delicacy, with not a drop spilled, marked a young doctor on the rise.

At the spa, pampered patrons had their blood let by roving bath attendants called *baglio men*. In the steam room, they'd be "cupped" by attendants who would scarify (cut) them with a lancet, apply a hot cup, and suck the blood out. At the customer's request, the resulting scar could be made into the shape of a heart or a love knot. In 1813, surgeons complained that bath attendants were cutting into their business.

Bloodletting was universal. Off the coast of Panama, British ship surgeon Lionel Wafer wrote in his diary:

The patient is seated on a stone in the river, and one with a small bow shoots little arrows into the naked body of the patient, up and down, shooting them as fast as they can ... if by chance they hit a vein which is full of wind, and the blood spurts out a little, they will leap and skip about, shewing many antic gestures, by way of rejoicing and triumph.

In the United States, doctors employed thumb lancets with tortoiseshell handles. The fleam, with blades at right angles to its elegant handle, was used on sick people, and pets and cattle as well.

The scarificator was the Cadillac of lancets. Suitable for placement next to the family Bible, in a fine wooden case, the scarificator was spring loaded, ivory trimmed, and mutton fat washable. It held up to twenty steel blades, each poised to flick outward at the click of a button.

On house calls, doctors would bring leeches with them in specially designed cases and jars with airholes.

Mysterious Episodes

Nothing could equal the horror of being buried alive. This happened with great frequency, according to doctors of the nineteenth century.

In 1819, the Doctor Regent of the Faculty of Medicine in Paris wrote, "One third, or perhaps half of those, who die in their beds, are not actually dead when they are buried." The *Lancet* called for "careful investigation" of these "mysterious episodes," and concerned doctors joined the London Association for the Prevention of Premature Burial.

Their concern may have been warranted—until the neighbors started complaining, one could never be sure when a body was actually dead. The *British Medical Journal* declared, "[H]ardly any one sign of death, short of putrefaction, can be relied upon as infallible." To try to elicit a reflex, doctors would stick objects down a body's throat, or tickle it with a feather, or blare a trumpet in its ear.

In 1822, Dr. Adolf Gutsmuth designed a "security coffin." Gutsmuth's coffin had a long tube connecting it to an aboveground monitoring station and an alarm to be triggered when the buried person finally awoke. Presumably the newly revived, though well rested, would be hungry and thirsty, so while waiting to be dug up, food and drink would be slid down the tube. Dr. Gutsmuth once buried himself in the coffin for an hour. He later repeated the stunt, this time for several hours, and, via the tube, enjoyed a repast of soup, beer, and sausages.

Waiting mortuaries, for that sensitive time between alleged death and actual decomposition, assured loved ones they were doing everything possible for their possibly deceased. They were built in cities all over Europe, some publicly funded. The best had teams of watchmen and ready supplies of food, drink, and, for those special occasions, cigars.

In the 1880s, Mark Twain visited a waiting mortuary:

It was a gristly place, this spacious room. There were 36 corpses of adults, stretched on their backs on slightly slanted boards . . . all of them with wax-white, rigid faces . . . Along the sides of the room . . . [were] several marble-visaged babies . . . Around a finger of each of these fifty still forms, both great and small, was a ring, and from the ring a wire led to the ceiling, and thence to a bell in a watch-room yonder, where, day and night, a watchman always sits alert and ready to spring.

The Death of James Garfield

In 1865, British surgeon Joseph Lister decided to keep himself and his surgeries clean, and his patients lived instead of died. Lister was celebrated and had a mouthwash named after him.

To American doctors, however, Lister was an object of ridicule—a finicky, neurotic crank. So was fussy Frenchman Louis Pasteur and his "invisible germs." Real medical men wore lab coats coated with blood and pus, and carried with them that good old "surgical stink."

In 1881, Charles J. Guiteau shot President James A. Garfield in the chest. Garfield was laid on a dirty floor and attended to by ten doctors in an hour. The president asked one of them what his odds were. The helpful doctor said, "One chance in a hundred."

Dr. Townsend stuck his unwashed finger into the wound until the arrival of the eminent Dr. Bliss, who took charge. Dr. Bliss used an unsterilized pipe as a probe. Dr. Purvis, the first black doctor to treat a president, told Bliss he was only making things worse. Bliss continued his poking and prodding.

Garfield was brought to the White House, where Surgeon General Wales and Dr. Woodward stuck their own dirty fingers into the wound. A surgeon from the Midwest dispatched an urgent letter to Garfield's wife. It said, "Do not allow probing of the wound. Probing generally does more harm than the ball . . . God help you."

As it turned out, the bullet wound was not as bad as the doctors had thought, and President Garfield began to recover. Dr. Bliss took the president off his usual diet and had him eat rich, fatty foods, such as bacon and lamb chops.

Weeks later, Garfield became sick again. His body was riddled with pus. Doctors again poked and prodded without washing themselves. When Dr. Bliss cut his hand on an instrument, he caught the president's infection and wore his hand in a sling. A pus sac near Garfield's eye became so big that when it ruptured, it flooded down Garfield's throat and nearly drowned him.

President Garfield died. An autopsy revealed that most of the tunneling wounds in his chest were caused not by the path of the bullet but by his doctors' fingers.

Several pints of pus were found pooled in Garfield's chest. It was this reservoir and the infection that caused it that caused him to die. One prominent physician, a disciple of Dr. Lister, claimed that Garfield would have survived if only he had been a regular guy, not a president. Then, the doctor said, Garfield would have been stitched up and left alone to heal, with no dirty fingers to make things worse.

Dr. Bliss rendered to Congress a "Statement of Services Rendered" for $25,000, the equivalent of more than half a million dollars today. Dr. Bliss said he deserved the goodwill of "every patriotic citizen" for his "great skill" and "the perfection of [his] surgical management."

Gone Fishing

TAPE WORM—REMOVED ALIVE IN TWO HOURS with HEAD or no charge (No fee in advance). Have cured over 2,000 people of tapeworms with this harmless, infallible specific, 50% of which were doctoring for other specific diseases, thereby eking out a miserable existence as thousands are doing. (Also cured two people of LIZARDS) Dr. J. G. Shipley, Montezuma, Iowa

—ADVERTISEMENT IN THE *BURLINGTON HAWKEYE*,
JULY 9, 1895

In 1855, *Scientific American* ran an article recognizing Dr. Alpheus Myers's contribution to internal medicine. Myers had invented a "Trap for Removing Tapeworms from the Stomach and Intestines."

The patient would fast for a week, to make sure the tapeworm was hungry. The trap would then be baited with cheese and lowered down the patient's throat with a string. The tapeworm, which the article claimed could be a hundred feet long, would take the bait. The trap would spring, and the worm would be reeled in.

According to *Scientific American*, the trap worked at least once, on a fifty-footer.

Humbugs of All Kinds

Harvard Medical School's Dr. Charles Oleson was a self-styled foe of "humbugs of all kinds." "[S]o often duped by these disgraceful quacks," Dr. Oleson wrote, the public deserved better.

Ayer's Vitanuova addiction tonic, Dr. Oleson discovered, contained both alcohol and cocaine. The Brinkerhoff System for Treating Piles (U.S. Patent #241,288) was so painful that patients threatened to sue. Recamier Moth and Freckle Syrup contained corrosive sublimate; Kickapoo Indian Oil, turpentine; Kennkle's Vegetable Worm Soup, dandelion oil.

In 1891, the crusading Dr. Oleson published *Secret Nostrums and Systems of Medicine*, a drug directory compiled from the "brightest and most practical of our medical journals."

Based on examination and laboratory analysis, Oleson was able to lend his considerable authority to Cram's Fluid Lightning, Wolcott's Pain Paint, Hamlin's Wizard Oil, and Centaur Liniment (For Man and for Beast). To cure baldness, Dr. Oleson offered a choice between Allen's World Hair Restorer, Ayer's Hair Vigor, and Dr. B.W.'s Hair Asthma Cure. For relief from sore eyes, whooping cough, and deafness, he recommended the Carbolic Smoke Ball.

Dr. Oleson became a trusted medical name, and his catalog was a huge success.

A Map of the Empire

About as useful as asking a patient to wiggle his ears, "mapping" a person's tongue became standard medical procedure for more than a hundred years.

Until 1843, the tongue was just a nuisance that blocked a good view of the mouth and throat. This changed when a medical journal grandly proclaimed that the tongue, through "The Doctrine of Sympathy," would give doctors nothing less than a "map of the empire of disease." All doctors had to do was know the difference between dotted, stippled, furry, and shaggy.

A year later, in 1844, Dr. Benjamin Ridge published his masterwork, *Glossology*. The sides of the tongue belonged to the kidneys, the tip to the intestines, and the edges to the brain. Of a sick person's tongue, Ridge wrote, "Its first pilous appearance resembles fine velvet; then coarse velvet; and in others . . . it resembles a spaniel's back just out of the water."

The tongue was also said to be a near-infallible lie detector. Its coating, or "fur," exposed patients who fudged medical histories and lied about their symptoms. Of conniving patients who tampered with their tongues, Ridge wrote, "[T]he Glossologist is not to be deceived by any scrapings . . . The laws of Nature . . . are not to be counteracted by so simple a manipulation."

Doctors now use a tongue depressor to push the tongue out of the way during their exams.

Lasting Impressions

In nineteenth-century Paris, pregnant women visited the Louvre to gaze at portraits of the comely and well proportioned. The women hoped the subjects' good looks would shape the fates of their own babies-to-be.

Discussed and debated in hundreds of medical journals, the doctrine of maternal impressions remained strong as ever. Karl Ernst von Baer, the father of embryology, embraced it, as did many, if not most, obstetricians.

Leading journals reported that in 1880, a pregnant woman was kicked in the head while milking a cow. She gave birth to a bovine-looking child. In 1889, a hungry cat gnawed two paws off a pet rabbit. The hare's terrified owner delivered a child with bad feet. In 1891, a farmer found a rabbit in the hay and threw it playfully at his expectant wife. No one laughed when their daughter was born with a mole on her face, containing fuzz like rabbit fur.

Dr. Lowman of South Carolina treated a pregnant woman who saw a buzzard eat a pig. Weeks later the woman miscarried and expelled a fetus that looked like a plucked bird, with wings. In 1889, the fetus was exhibited before the New York Pathological Society at its gala anniversary meeting. The phenomenon wasn't limited to people. In 1899, the *British Medical Journal* reported that a sensitive dog suffering from a broken leg gave birth to a pup with a club foot.

Most persuasive of all, in 1898 the *American Journal of Obstetrics* presented the case of a pregnant woman who exhibited a craving for sunfish. As a surprise, her adoring husband brought back his still-wriggling catch in a pail and placed the pail on the front porch. Photographs published in the journal showed the couple's otherwise lovely daughter with a fish-shaped spot on her leg, and accompanying text noted that the girl herself enjoyed eating sunfish.

In 1903, the *American Textbook of Obstetrics* endorsed the doctrine of maternal impressions, citing as "familiar evidence of this anomaly" the "well-known 'elephant man' of England and the 'turtle man' exhibited in the United States."

Letter to the Editor

In 1867, the *Bulletin of the New York Academy of Medicine* published a letter by Dr. A. C. Post, recounting his heroic feats in bloodletting.

CASE I

A gentlemen of this city . . . had overtaxed his brain . . . I opened a vein in the arm, and bled him copiously. [He] subsequently had several attacks . . . each time [he] was relieved from them by the abstraction of blood. At a later period he suffered occasional attacks of coma . . . Local bloodletting was then resorted to.

CASE II

A young clergyman, to whom I was called to see . . . the flow of blood was followed by marked and complete relief. The pressure on the brain was so severe that, if speedy relief had not been afforded . . . the patient would have been permanently incapacitated . . .

*I was induced to bleed her at her own urgent request . . .
I was called during three or four similar attacks . . . after
midnight, [I] found her in a heavy stupor—unable to
speak. She, however, recognized me, and without being
able to speak, pointed to her elbow, indicating that I
should bleed her; which I did.*

In another case, I believe, I saved two lives . . .

Spring Cleaning

Like many Victorians, Sir Arbuthnot Lane was preoccupied with his bowels.

Considered England's finest abdominal surgeon and knighted by the queen in 1913, Lane had two problems with his intestines. First, they were folded in on themselves—"kinked," in his words. Worse, they were "a cesspool," a cesspool that needed to be drained with "regular and systemic flushing." After discussing the matter with colleagues at a dinner party, he had his family drink two gallons of paraffin to grease their intestines. He later required the same of his servants and grandkids, and his pet monkey.

Adults complained that the oily substance made them pass gas and leaked onto their clothes. One woman presented her doctor with a dry-cleaning bill. Children were teased by their classmates.

Dr. Lane designed the girdlelike "Curtis belt" to provide support and iron out the kinks. When kids complained about the belt's awkward appearance, a colleague of Dr. Lane said that, eventually, they would "[f]ace the ridicule of their schoolfellows rather than give it up."

The paraffin and the belts were a good start, but they weren't nearly enough. Lane was destined, he felt, to personally stamp out, once and for all, the "cause of all the chronic diseases of civilization." And Dr. Lane brimmed with confidence. An American doctor said of him, "I wish I was as sure of anything as [he] is of everything."

Dr. Lane began removing small sections of his patients' dirty intestines. The procedure helped, a little. Then came the breakthrough.

Dr. Lane read the work of Nobel Prize–winning biologist Elie Metchnikoff, a deep thinker who had recently discovered that because of evolution, our large intestine was essentially obsolete. Alarmed at the prospect of bacteria multiplying at "128,000,000,000,000 each day," Metchnikoff proposed that the intestine, doomed to shrink anyway, be taken out immediately, like a bad appendix. In support of his theory, Metchnikoff noted that parrots, whose intestines were very small, lived long, happy lives.

Emboldened, Dr. Lane took to removing his patients' entire large intestines. He took out more than a thousand.

The Medical Standard applauded. Appalled that "ileal torsion" (kinks) and the like had caused an "invasion of the microbes," especially among members of Guy's football team in London, the journal recommended use of the Curtis belt and also a vaccine composed of urine and feces. However:

> [W]hen the large intestine has become diseased beyond
> repair by the ravages of microbes ... the best prospect is
> undoubtedly found in the removal of the large intes-
> tine ... the results of a successful colectomy are so bril-
> liant as to seem almost miraculous.

Some doctors, perhaps less familiar with the works of the great Metchnikoff, thought Lane went too far. Of the theory behind his surgeries, one doctor quipped, "A kink to you and me is no kink to the colon."

But Dr. Lane was feted on both sides of the Atlantic. In 1912, he was an honored speaker at the Clinical Congress of Surgeons in New York, which also featured surgical demon- strations conducted on mentally defective children taken from a nearby public school. One journal looked forward to a future race with "foreshortened colons," and a fellow surgeon suggested that while removing the intestines, a doctor might also wish to do some "spring cleaning" of other organs as well.

The surgery eventually lost its luster, even to Lane. He repudiated his theory in 1926, and became an advocate of exercise and healthy eating.

The Foot-O-Scope

In 1895, Wilhelm Roentgen took the world's first x-ray. It showed the bones of his wife's hand, and radiology was born. Boston physician Jacob Lowe had another great idea, and in 1927 received a patent for his own invention, the Foot-O-Scope. It helped people try on shoes.

Harnessing the mysterious power of the x-ray was thought necessary to properly "palpate" the foot and toes and to avoid "deform[ing] the sensitive bone joints." Assistance from an "advisory friend" and a shoe salesman was also recommended. The sophisticated technology received the *Parents Magazine* Seal of Approval.

The fit was fine, but customers mostly liked the "shoe fluoroscope" because it was fun. For the first time ever, kids wanted to be in a shoe store. Store owners liked it too—they bought ten thousand of them. The dishwasher-sized units cost $2,000 apiece, a small fortune at the time.

A child would slide his foot into the machine's base, while parents, salesmen, and beaming siblings watched the action through specially designed vents and windows. Each press of the button triggered an x-ray cycle up to twenty seconds long. People loved to press that button, and the more presses, the better the fit.

At around the same time, the Tricho machine made its debut. Installed alongside hair dryers in beauty salons, the device was invented by Dr. Geyser of New York, a retired professor of physiological therapy at Fordham University. According to the marketing material, Dr. Geyser had devoted his entire life to removing hair from women's lips, and with his patented x-ray beam had finally succeeded.

Years later, with atom bombs going off, doctors examined the fluoroscope's radiation levels. The bulky devices emitted way more radiation than regular x-ray machines, and their particles shot out twenty-five feet in any direction. Depending on the machine, a little girl trying on just a few pairs of shoes could receive enough radiation to have her growth stunted. Store clerks, exposed on a daily basis, presumably just died off, though no one bothered to check.

Radiation turned out to be a bigger threat than tight-fitting shoes, and federal authorities banned the Foot-O-Scope and the Pedoscope in 1953.

The Tooth Fairy

Dr. Henry Cotton had finally discovered the cause of mental illness—bad teeth. By yanking a rotten molar, a doctor could stop "focal infection" in its tracks and rescue its owner from the ravages of depression, schizophrenia, and psychosis.

By the 1920s, focal infection theory, more scientific than Freud and his ilk, seemed on the verge of revolutionizing medicine, and maybe the entire future course of human history as well. All a doctor had to do was pull out an infected tooth and, perhaps, any tooth that might go bad some time later. At Trenton State Hospital, Dr. Cotton and his staff pulled out more than eleven thousand teeth, including some of Dr. Cotton's own and those of his wife and children.

It wasn't just teeth that got infected, so Dr. Cotton began removing his patients' organs as well. Many, Dr. Cotton happily discovered, were really quite expendable. Dr. Cotton took out kidneys, gall bladders, ovaries, and just about anything else he could get his hands on. Few complained, and the ones who did were considered crazy, so no one listened.

Dr. Cotton's mental patients, now clean as a whistle, made remarkable recoveries, and his epic findings were confirmed by prominent doctors all over the world. In 1922, the *New York Times* wrote:

> *At the State Hospital at Trenton, N.J., under the brilliant leadership of the medical director, Dr. Henry A. Cotton, there is on foot the most searching, aggressive, and profound scientific investigation that has yet been made of the whole field of mental and nervous disorders . . . there is hope, high hope . . . for the future.*

Dr. Cotton's patients weren't getting better; they were dying, many on the operating table. Dr. Cotton himself began to act "peculiar." But an investigation was squelched, and the organ removals, and deaths, continued.

Dapper and well connected, Dr. Cotton remained the toast of the medical establishment. Medical journals lauded him, and dinners were held in his honor. He was cited for his "rare surgical judgment and technique" at "the most progressive institution in the world for the care of the insane." Wealthy people lined up to have their own teeth taken out at expensive private retreats and spas.

In 1925, Dr. Cotton became mentally ill himself. After taking some rest and having a few more teeth taken out, he went back to work.

When Dr. Cotton died, the *American Journal of Psychiatry* wrote of the "[e]xtraordinary record by one of the most stimulating figures of our generation," and the *New England Journal of Medicine* hailed Dr. Cotton's great contributions to patient health and well-being.

Anchors Away

Loosened from their bodily moorings, "floating kidneys" were often found in upper-class women, especially those said to be "pinched-looking" or of a nervous and excitable disposition. Many doctors found the "condition" hard to diagnose, or maybe even nonexistent, but some surgeons detected them easily, usually by palpating the abdomen and evaluating the woman's "countenance."

Floating kidneys could be caused by tripping over a high step, or being pushed from a streetcar, or falling from a horse, or wearing a tight corset. Too much dancing or a bad cough could cause a person to drop a kidney. To treat the problem, surgeons would anchor the kidneys to adjoining muscle. The procedure had its risks but was much in demand, especially among the upper crust.

Leading medical journals warned that patients with kidneys adrift were doomed to "a life of greater or lesser suffering." If the condition was left untreated, symptoms included memory loss, sexual difficulties, and insanity. Suicide was not unknown.

The Glasgow Royal Infirmary did almost 250 procedures, Dr. Goelet of New York City more than a hundred just by himself. Goelet also lectured other surgeons so they could do it. London's leading chest surgeon, Sir Arbuthnot Lane, did several.

In 1905, British surgeon Cornelius Suckling, author of *Movable Kidney: A Cause of Insanity, Headache, Neurasthenia, Insomnia, Mental Failure and Other Disorders of the Nervous System* (1905), palpated the kidneys of fifty women produced for his inspection at a lunatic asylum. An astonishing 58 percent suffered from kidney drop, the diagnosis confirmed by the patients' drawn faces and worried looks.

As a prophylactic, Dr. Suckling supported the use of kidney belts, especially the inflatable kind. He wrote of one woman who threatened to slit her throat when her belt became worn but felt better after buying a new one at the store. But surgery was easier and faster, and Dr. Suckling performed dozens of operations. All were apparently successful.

One husband said his wife had formerly been dazed and stupid, but after her tightening she was able to do housework and pick fruit. Another woman no longer thought her head was cut off. Still another woman became busy and cheerful and knitted stockings.

Shock Therapy

Shock therapy didn't cure patients, but it did make them act nice.

Its pioneer was Ugo Cerletti, who complained to colleagues that using the electric chair for executions was giving his field a bad name. One day Cerletti went to a slaughterhouse to see how pigs were killed. Seeing them dispatched with a simple jolt to the head, he left inspired.

Luckily enough, a thirty-nine-year-old homeless person, rounded up by the local police, was delivered right to Cerletti's door. Cerletti gave him 80 volts, then 90. Nothing happened, except that the patient broke into song. As Cerletti prepared to administer more voltage, the patient screamed, "*Non una seconda! Mortifera!*" (Not a second time! It will kill me!) Cerletti zapped him with 110 volts and induced a seizure. It was a breakthrough.

Later, Cerletti injected patients with brain matter from electrified pigs (electroshock pig brain therapy) and tried to stimulate his patients' "vitalizing substances," which he called *agro-agonines*, Greek for "extreme struggle." Cerletti also designed army uniforms and fuses for artillery shells. In 1950, he was nominated for the Nobel Prize, though not the Nobel Peace Prize.

Electroshock therapy was considered a boon for bickering families and romances gone sour. In 1959, the parents of seventeen-year-old Jonika Upton committed her to a mental hospital. She had run off with an artist boyfriend and, before that, dated a boy with suspected homosexual tendencies. According to medical records, Jonika also "walked about carrying Proust under her arm."

After sixty-two separate shocks, Jonika's prognosis remained bleak. "[S]he has not become nearly foggy as we might wish," her doctors wrote. Eventually, however, she began to display encouraging signs of "dilapidation," like incontinence, and was seen walking around the grounds naked and Proust-free. She had forgotten all about that artsy boyfriend.

Today, electroshock therapy, applied under strictly controlled circumstances, is considered a legitimate treatment for the mentally ill.

Eureka! (Not)

Schoolchildren know the amazing story of penicillin. In 1929, a carelessly placed petri dish was accidentally contaminated with a miracle mold that routed the deadly germs around it. A wonder drug was born, and millions were saved.

Penicillin was a wonder drug, but no thanks to its famous discoverer, Dr. Alexander Fleming. A man not easily impressed, Fleming was so underwhelmed by his bacteria-eating prodigy that he kept it on the shelf—for ten years.

A few months after his epic find, Fleming did get around to hiring a couple of assistants to isolate penicillin's magic ingredient, and they came very close. But one quit for a better job, and the other went on a cruise. Later, some lab workers tried to refine the active compound, but they gave up after a few weeks.

Sometimes Fleming's thoughts would drift back to the miracle mold he'd discovered, potential savior of millions, and at times like this he'd . . . complain. If only his boss at the lab were more supportive, he'd grouse to colleagues, maybe things would finally get done. The thought would be fleeting, however, and Fleming would get back to work—on something else.

Ten years later, as Fleming puttered in his lab, two more energized scientists, Howard Florey and Ernst Boris Chain, picked up where he had left off and quickly developed the wonder drug. Fleming didn't lift a finger until their breakthrough became evident, at which point he decided to honor penicillin's real heroes with a lab visit. Unimpressed with both Fleming and his contributions, Chain remarked, "Good God, I thought he was dead."

All three received the Nobel Prize in 1945, but it's Fleming's name that lives on.

Born Again

Administering huge doses of insulin, psychiatrists after World War II plunged patients into comas, then brought them back. When the effects wore off, the doctors would do it again, and again after that. One patient was "treated" and revived sixty times in two months.

The insulin would deprive the brain of its fuel and kill a certain percentage of brain cells. Through this "cleansing" of the brain, the patient would experience a "loss of tension and hostility."

The treatment worked, in a sense. Formerly difficult patients would be transformed into smiling little children, demanding nothing more than to be hugged, fed, and showered with balloons and teddy bears. The sudden reversal was a godsend to beleaguered families wanting to express their love and desperate for a cure, and a boon to doctors and nurses used to walking grim psychiatry wards.

According to Dr. Manfred Sakel, the procedure's inventor:

An adult patient, for example, will say . . . he is six years old. His entire behavior will be childish . . . the timbre of

his voice and his intonation are absolutely infantile . . .
He asks in a childish, peevish way when he may go to
school. He says he has a tummy ache.

This happy circumstance would last until the patient reverted to the troubled adult he always was, and then the cycle would begin anew.

Decades later, researchers finally got around to studying their miracle technique and realized that putting someone in a coma was no more effective than hitting them on the head with a hammer. One disgusted physician said, "It gave them [psychiatrists] something to do. It made them feel like real doctors . . ."

The Uline Ice Company

You might as well talk about a successful automobile accident.

—FRANKLIN FREEMAN, SON OF WALTER FREEMAN

Nothing worked for mental illness. Doctors were desperate for a cure.

In 1939, Egas Moniz of Portugal drilled holes in the skulls of mentally ill people and destroyed their brain tissue with wires and alcohol. The theory behind the technique remained murky, but it earned him a Nobel Prize. Moniz had a vested interest in treating mental illness—a disturbed patient had shot and paralyzed him. "That crazy man plugged me with bullets," Moniz said.

One of Moniz's protégés was Walter Freeman, from Yale. Dr. Freeman was not a surgeon, and he got frustrated when his instruments would break off in patients' skulls. Dr. Freeman had neither the time nor the patience to wear gloves or create a sterile field—"all that germ crap," he called it. At home, on a grapefruit, Freeman began practicing a new, easier technique.

In 1946, Freeman unveiled his new procedure, the transorbital lobotomy. Instead of wasting time with the skull, Freeman would go right through the eye. In the beginning Freeman generally used ice picks from his kitchen drawer, especially those from the Uline Ice Company. He'd hammer them in with a mallet, and then scrape, like a windshield wiper. Attending doctors often fainted.

In front of medical students, Freeman liked to show off, writing on a chalkboard with both hands at the same time. He began doing this with his patients, going through both eyes at once. This impressed reporters and made things go faster.

During the summer, Freeman would pack his wife, kids, and surgical tools into the family van, called the "lobotomobile," and perform the procedure at local hospitals while on his way to the national parks.

One day, while at home, Dr. Freeman was alerted that a patient had barricaded himself in his bedroom. Freeman got dressed, went to the scene, and talked the patient out of his room. He then had a policeman hold the patient down and did his surgery on the floor. At Cherokee State hospital in Iowa, three of Freeman's patients expired. One died because the ice pick Freeman was using slipped while he was taking a photograph.

In medical charts and records, Freeman freely noted

that many of his patients now functioned at "household pet" level or turned into "good solid cake but no icing." But they became solid citizens and gave no trouble.

As long-term studies became available and the medical tide turned against him, Freeman became baffled and deeply hurt. Many patients loved him, and their families loved him even more. When Freeman was heckled at a meeting of psychiatrists, he grabbed a box from under the lectern, dumped five hundred Christmas cards onto the table, and asked how many holiday greetings from patients his audience had received that year.

After Freeman botched yet another procedure, he was forced to retire. He packed his bags and traveled the country, tracking down old patients to see how they were doing. He would show the other doctors, and himself, that he was a good man, who only wanted to help.

Selected Bibliography

Scholarly Debate About the Course of Medical "Progress"

Wootton, David. "Bad Medicine: Doctors Doing Harm Since Hippocrates." www.badmedicine.co.uk/main.asp.

"Twisted" Medical History (by an Oxford-Trained Medical Historian)

Fitzharris, Lindsey. *The Chirurgeon's Apprentice* (blog). http://thechirurgeonsapprentice.com.

General Medical History

Adler, Robert E. *Medical Firsts: From Hippocrates to the Human Genome.* Hoboken, NJ: Wiley, 2004.

Bynum, W. F., and Roy Porter. *Companion Encyclopedia of the History of Medicine.* London: Routledge, 1993.

Clendening, Logan. *Source Book of Medical History.* New York: Dover, 1960.

Eamon, William. *The Professor of Secrets: Mystery, Medicine, and Alchemy in Renaissance Italy.* Washington, DC: National Geographic, 2010.

Gordon, Richard. *The Alarming History of Medicine.* New York: St. Martin's Press, 1994.

Kelly, Kate. *The History of Medicine.* New York: Facts on File, 2009.

Lehrer, Steven. *Explorers of the Body*. Garden City, NY: Doubleday, 1979.

Majno, Guido. *The Healing Hand: Man and Wound in the Ancient World*. Cambridge, MA: Harvard University Press, 1975.

Porter, Roy. *Blood and Guts: A Short History of Medicine*. New York: Norton, 2003.

———. *The Greatest Benefit to Mankind: A Medical History of Humanity*. New York: Norton, 1998.

———. *Medicine: A History of Healing*. London: O'Mara, 1997.

Reiser, Stanley Joel. *Medicine and the Reign of Technology*. Cambridge, UK: Cambridge University Press, 1978.

———. *Technological Medicine: The Changing World of Doctors and Patients*. New York: Cambridge University Press, 2009.

Sigerist, Henry E. *The Great Doctors: A Biographical History of Medicine*. Garden City, NY: Doubleday, 1958.

———. *A History of Medicine*. New York: Oxford University Press, 1951.

Thorndike, Lynn. *A History of Magic and Experimental Science*. New York: Columbia University Press, 1958.

Wootton, David. *Bad Medicine: Doctors Doing Harm Since Hippocrates*. Oxford, UK: Oxford University Press, 2006.

Wynbrandt, James. *The Excruciating History of Dentistry*. New York: St. Martin's Press, 1998.

Specific Medical Topics

CHAPTER ONE

Babylonia

Adamson, P. B. "Surgery in Ancient Mesopotamia." *Medical History* 35 (1991): 428–435.

Halsall, Paul. "Internet Ancient History Sourcebook." Fordham University. www.fordham.edu/halsall/ancient/asbook.asp.

Sasson, Jack M. *Civilizations of the Ancient Near East*. New York: Scribner, 1995.

Egypt

Estes, J. Worth. *The Medical Skills of Ancient Egypt.* Canton, MA: Science History Publications/USA, 1989.

Middendorp, Joost, Gonzalo Sanchez, and Alwyn Burridge. "The Edwin Smith Papyrus: A Clinical Reappraisal of the Oldest Known Document on Spinal Injuries." *European Spine Journal,* November 19 (2010): 1815–1823.

Greece

Adams, Francis, ed. *The Genuine Works of Hippocrates.* Huntington, NY: Krieger, 1972.

Gilman, Sander L. *Hysteria Beyond Freud.* Berkeley: University of California Press, 1993.

King, Helen. *Hippocrates' Woman: Reading the Female Body in Ancient Greece.* London: Routledge, 1998.

Lefkowitz, Mary R., and Maureen B. Fant. *Women's Life in Greece and Rome.* Baltimore: Johns Hopkins University Press, 1982.

Lloyd, G. E. R., ed. *Hippocratic Writings.* Harmondsworth, NY: Penguin, 1978.

Roccatagliata, Giuseppe. *A History of Ancient Psychiatry.* New York: Greenwood Press, 1986.

Rome

"Etruscan and Roman Medicine." Claude Moore Health Sciences Library. www.hsl.virginia.edu/historical/.

King, Helen. *Greek and Roman Medicine.* London: Bristol Classical Press, 2001.

CHAPTER TWO

Medieval Medicine—General

Getz, Faye Marie. *Medicine in the English Middle Ages.* Princeton, NJ: Princeton University Press, 1998.

Lacey, Robert, and Danny Danziger. *The Year 1000: What Life Was Like at the Turn of the First Millennium: An Englishman's World.* Boston: Little, Brown, 1999.

McVaugh, M. R. *Medicine Before the Plague: Practitioners and Their Patients in the Crown of Aragon, 1285–1345.* Cambridge, UK: Cambridge University Press, 1993.

Pouchelle, Marie-Christine. *The Body and Surgery in the Middle Ages.* New Brunswick, NJ: Rutgers University Press, 1990.

Prioreschi, Plinio. *Medieval Medicine.* Omaha, NE: Horatius Press, 2003.

Rawcliffe, Carole. *Medicine and Society in Later Medieval England.* Stroud, UK: Sutton, 1995.

Sigerist, Henry E. *Medieval Medicine.* Philadelphia: University of Pennsylvania Press, 1941.

Siraisi, Nancy G. *Medieval and Early Renaissance Medicine: An Introduction to Knowledge and Practice.* Chicago: University of Chicago Press, 1990.

Tuchman, Barbara Wertheim. *A Distant Mirror: The Calamitous 14th Century.* New York: Knopf, 1978.

Wallis, Faith. *Medieval Medicine: A Reader.* Toronto: University of Toronto Press, 2010.

Walsh, James J. *Medieval Medicine.* London: Black, 1920.

Walsh, James J., and Leon Banov. *Old-Time Makers of Medicine: The Story of the Students and Teachers of the Sciences Related to Medicine During the Middle Ages.* New York: Fordham University Press, 1911.

Astrology

Bullough, Vern L., Marie-Christine Pouchelle, and Rosemary Morris. "The Body and Surgery in the Middle Ages." *American Historical Review* 97, no. 1 (1992): 177. doi:10.2307/2164569

Surgery and Surgeons

Daremberg, Charles V., ed. *The Surgery of Roland of Parma* (trans. Leonard D. Rosenman). San Francisco: Xlibris, 2001.

De Mets, A., and Mario Tabanelli, ed. *The Surgery of Master Jehan Yperman (1260?–1330?)* (trans. Leonard D. Rosenman). Philadelphia: Xlibris, 2002.

Ismail, Anis, and A. B. Khan. "Surgery in the Medieval Muslim World." *Surgery in the Medieval Muslim World* 19, no. 1 (1964): 64–70.

McVaugh, M. R. *The Rational Surgery of the Middle Ages*. Florence, Italy: SISMEL/Edizioni del Galluzzo, 2006.

Nicaise, E., Jean Baptiste Saint-Lager, and F. Chavannes, eds. *The Surgery of Master Henri de Mondeville: Written from 1306 to 1320* (trans. Leonard D. Rosenman). Philadelphia: Xlibris, 2003.

Pifteau, Paul, ed. *The Surgery of William of Saliceto: Written in 1275* (trans. Leonard D. Rosenman). Philadelphia: Xlibris, 2002.

Rosenman, Leonard D., trans. *The Chirurgia of Roger Frugard*. Philadelphia: Xlibris, 2002.

Tabanelli, Mario. *The Surgery of Bruno da Longoburgo: An Italian Surgeon of the Thirteenth Century*. Pittsburgh, PA: Dorrance, 2003.

von Fleischhacker, Robert, ed. *The Surgery of Lanfranchi of Milan: A Modern English Translation* (trans. Leonard D. Rosenman). Philadelphia: Xlibris, 2003.

John of Gaddesden

Cholmeley, Henry Patrick. *John of Gaddesden and the Rosa Medicinae*. Oxford, UK: Clarendon Press, 1912.

Rawcliffe, Carole. *Medicine and Society in Later Medieval England*. Stroud, UK: Sutton, 1995.

Bald's Leechbook

Lacey, Robert, and Danny Danziger. *The Year 1000: What Life Was Like at the Turn of the First Millennium: An Englishman's World*. Boston: Little, Brown, 1999.

Wright, C. E., ed. *Bald's Leechbook*. Copenhagen: Rosenkilde and Bagger, 1955.

Urology

Harvey, Ruth. "The Judgment of Urines." *Canadian Medical Association Journal* 159, no. 12 (1998): 1482–1484.

Bedside Manner

Linden, David E. J. "Gabriele Zerbi's *De Cautelis Medicorum* and the Tradition of Medical Prudence." *Bulletin of the History of Medicine* 73, no. 1 (1999): 19–37. doi:10.1353/bhm.1999.0035

Medieval Dentistry

Anderson, T. "Dental Treatment in Medieval England." *British Dental Journal* 197 (2004): 419–425.

Hildegard of Bingen

Maddocks, Fiona. *Hildegard of Bingen: The Woman of Her Age.* New York: Doubleday, 2001.

Strehlow, Wighard, and Gottfried Hertzka. *Hildegard of Bingen's Medicine.* Santa Fe, NM: Bear, 1988.

The Trotula

Green, Monica H., ed. *The Trotula: A Medieval Compendium of Women's Medicine.* Philadelphia: University of Pennsylvania Press, 2001.

Bad Doctor

Halsall, Paul. "Internet Ancient History Sourcebook." Fordham University. www.fordham.edu/halsall/ancient/asbook.asp.

CHAPTER THREE

King Charles

"Source Analysis—Death of Charles II Based on Scarburgh's Description." BBC News. www.bbc.co.uk/schools/gcsebitesize/.

Beak Doctors

"Elizabethan Medicine and Illnesses." www.elizabethan-era.org .uk/elizabethan-medicine-and-illnesses.htm.

Fitzharris, Lindsey. "Behind the Mask: The Plague Doctor." *The Chirurgeon's Apprentice.* March 13, 2012. http://thechirurgeon sapprentice.com/2012/03/13/behind-the-mask-the-plague-doctor/.

Weapon Salve and Sympathetic Powder

Hedrick, Elizabeth. "Romancing the Salve: Sir Kenelm Digby and the Powder of Sympathy." *British Journal for the History of Science* 41, no. 2 (2008). doi:10.1017/S0007087407000386

Osler, Sir William. *Sir Kenelm Digby's Powder of Sympathy: An Unfinished Essay.* Los Angeles: Plantin Press, 1972.

Paracelsus

Lund, Fred B. "Paracelsus." *Annals of Surgery* 94, no. 4 (1931): 548–561. doi:10.1097/00000658-193110000-00009

Stillman, John Maxson. *Paracelsus*. New York: AMS Press, 1982.

Stoddart, Anna M. *The Life of Paracelsus: Theophrastus von Hohenheim, 1493–1541*. London: Rider, 1915.

Corpse Medicine

Bishop, W. J. *The Early History of Surgery*. London: Hale, 1960.

Fitzharris, Lindsey. "Drinking Blood and Eating Flesh: Corpse Medicine in Early Modern England." *The Chirurgeon's Apprentice*. February 25, 2011. http://thechirurgeonsapprentice.com/2011/02/25/drinking-blood-and-eating-flesh-corpse-medicine-in-early-modern-england/.

Lehrer, Steven. *Explorers of the Body*. Garden City, NY: Doubleday, 1979.

Sugg, Richard. "Corpse Medicine: Mummies, Cannibals, and Vampires." *The Lancet* 371, no. 9630 (2008): 2078–2079. doi:10.1016/S0140-6736(08)60907-1

Maternal Impressions

Bhattacharya, S., V. Khanna, and R. Kohli. "Cleft Lip: The Historical Perspective." *Indian Journal of Plastic Surgery* 42 (suppl.), October (2009): S4–S8.

Tubbs, W. J. "Influence of Mental Impressions on the Foetus in Utero." *Provincial Medical and Surgical Journal*, December 21 (1842): 268–269.

Vesalius

Castiglioni, Arturo. "The Attack of Franciscus Puteus on Andreas Vesalius and the Defence by Gabriel Cuneus." *Yale Journal of Biology and Medicine* 16 (December 1943): 135–148.

Doctrine of Signatures

Pearce, J. M. S. "The Doctrine of Signatures." *European Neurology* 60, no. 1 (2008): 51–52. doi:10.1159/000131714

Rafeeque, Muhammed. "The Doctrine of Signature." *The Homoeo-pathic Heritage*, May 2008.

Executioner Medicine

Stuart, Kathy. *Defiled Trades and Social Outcasts: Honor and Ritual Pollution in Early Modern Germany*. Oxford, UK: Cambridge University Press, 1999.

Lovesickness

Ferrand, Jacques, Donald Beecher, and Massimo Ciavolella. *A Treatise on Lovesickness*. Syracuse, NY: Syracuse University Press, 1990.

Nose Jobs

Grant, Edward. *A Sourcebook in Medieval Science*. Cambridge, MA: Harvard University Press, 1974.

Whitaker, Ian S. "Corrections to the Birth of Plastic Surgery." *Plastic and Reconstructive Surgery: Journal of the American Society of Plastic Surgery* 121, no. 3 (2008): 1072–1073.

Public Anatomy

Klestinec, C. "A History of Anatomy Theaters in Sixteenth-Century Padua." *Journal of the History of Medicine and Allied Sciences* 59, no. 3 (2004): 375–412. doi:10.1093/jhmas/59.3.375

Book of Medicines

Millard-Rosenberg, S. L. "Sixteenth Century German Medicine." *Western Journal of Medicine* 331, no. 1 (July 1930): 508–512.

Nostalgia

"Editorial: Nostalgia: A Vanished Disease." *British Medical Journal* 6014 (1976): 857–858. doi:10.1136/bmj.1.6014.857

Kiser-Anspach, Carolyn, trans. "Medical Dissertation on Nostalgia by Johannes Hofer, 1688." *Bulletin of the Institute of the History of Medicine* (1934): 376–391.

Syphilis

Hayden, Deborah. *Pox: Genius, Madness, and the Mysteries of Syphilis*. New York: Basic Books, 2003.

Quétel, Claude. *History of Syphilis.* Baltimore: Johns Hopkins University Press, 1990.

Bosum Serpents

Bondeson, Jan. "The Bosum Serpent." *Journal of the Royal Society of Medicine* 91 (August 1998): 442–447.

Pouchelle, Marie-Christine. *The Body and Surgery in the Middle Ages.* New Brunswick, NJ: Rutgers University Press, 1990.

A Cabinet of Curiosities

Bondeson, Jan. *A Cabinet of Medical Curiosities.* Ithaca, NY: Cornell University Press, 1997.

"Collecting Curiosities: The Rise of the Museum." New York Academy of Medicine. www.nyam.org/library/rare-book-room/exhibits/telling-of-wonders/ter6.html.

Sappol, Michael. "A Cabinet of Curiosities." Common-Place. www.common-place.org/vol-04/no-02/sappol/.

Nicholas Culpeper

Brockbank, William. "Sovereign Remedies: A Critical Depreciation of the 17th-Century London Pharmacopeia." *Medical History* 8, no. 1 (1964): 1–14.

Culpeper, Nicholas. *The English-Physicians Dayly Practise. Or, Culpeper's Faithful Physitian. Teaching Every Man and Woman to Be Their Own Doctor* . . . London: Printed for J. Conyers at the Black Raven in Duck Lane, 1680.

Woolley, Benjamin. *Heal Thyself: Nicholas Culpeper and the Seventeenth-Century Struggle to Bring Medicine to the People.* New York: HarperCollins, 2004.

Man-Midwives

Cassidy, Tina. *Birth: The Surprising History of How We Are Born.* New York: Atlantic Monthly Press, 2006.

Cook, James Wyatt, and Barbara Collier Cook. *Man-Midwife, Male Feminist: The Life and Times of George Macaulay, M.D., Ph.D. (1716–1766).* Ann Arbor: Scholarly Publishing Office, University of Michigan Library, 2004.

Male Jewish Menstruation

Beusterien, John L. "Jewish Male Menstruation in Seventeenth-Century Spain." *Bulletin of the History of Medicine* 73, no. 3 (1999): 447–456. doi:10.1353/bhm.1999.0097

Breast-Feeding

Sherwood, Joan. "The Milk Factor: The Ideology of Breastfeeding and Post-Partum Illnesses, 1750–1850." *Canadian Journal of Medical History* 10 (1993): 25–47.

The Secret

Dunn, Peter M. "The Chamberlen Family (1560–1728) and Obstetric Forceps." *Archives of Disease in Childhood: Fetal and Neonatal* 81, no. 3 (November 1999): 232–235.

The Microscope

Wolfe, Charles T. "Empiricist Heresies: The Polemic Against Experiment in Early Modern Medical Thought." Working paper. http://sydney.edu.au/science/hps/empiricism/downloads/Conference_2009/Embodied_Empiricism_09_Papers/Wolfe_Empiricist_heresies.pdf.

Wolfe, David E. "Sydenham and Locke on the Limits of Anatomy." *Bulletin of the History of Medicine* 25, no. 3 (May/June 1961): 193–220.

Warm Beer

F. W. *A Treatise of Warm Beer Wherein Is Declared by Many Reasons That Beer So Qualified Is Farre More Wholesome Then That Which Is Drunk Cold.* Cambridge, UK: Overton, 1641.

Witches

Gevitz, N. "'The Devil Hath Laughed at the Physicians': Witchcraft and Medical Practice in Seventeenth-Century New England." *Journal of the History of Medicine and Allied Sciences* 55, no. 1 (2000): 5–36.

"Witchcraft and Medicine." *The Lancet* 209, no. 5406 (1927): 769.

Zilboorg, Gregory. "The Medical Man and the Witch Towards the Close of the Sixteenth Century." *Bulletin of the New York Academy of Medicine* 11, no. 10 (1935): 579–607.

Tarantism

Bartholomew, Robert E. "Rethinking the Dancing Mania." Committee for Skeptical Inquiry. www.csicop.org/si/show/rethinking_the_dancing_mania.

Carlson, Eric T., and Meribeth M. Simpson. "Tarantism or Hysteria? An American Case of 1801." *Journal of the History of Medicine and Allied Sciences* 26, no. 3 (1971): 293–302. doi:10.1093/jhmas/XXVI.3.293

Russell, Jean. "Tarantism." *Medical History* 23, no. 4 (October 1979): 404–425.

Enemas

Rosenhek, Jackie. "Royal Flush." *Doctor's Review.* July 2005. www.doctorsreview.com/history/jul05-history/.

Zacks, Richard. *An Underground Education: The Unauthorized and Outrageous Supplement to Everything You Thought You Knew About Art, Sex, Business, Crime, Science, Medicine, and Other Fields of Human Knowledge.* New York: Doubleday, 1997.

The Bezoar

Paget, Stephen. *Ambroise Paré and His Times, 1510–1590.* New York: Putnam, 1897.

Spontaneous Generation

Bondeson, Jan. *The Feejee Mermaid and Other Essays in Natural and Unnatural History.* Ithaca, NY: Cornell University Press, 1999.

CHAPTER FOUR

Benjamin Rush

Rush, Benjamin. *Medical Inquiries and Observations, Upon the Diseases of the Mind.* Philadelphia: Johnson and Warner, 1812.

Anesthesia

Fenster, J. M. *Ether Day: The Strange Tale of America's Greatest Medical Discovery and the Haunted Men Who Made It*. New York: HarperCollins, 2001.

Centrifuge Therapy

Harsch, Viktor. "'Centrifuge Therapy' for Psychiatric Patients in Germany in the Early 1800s." *Aviation, Space and Environmental Medicine* 77 (2006): 157–160.

Wade, N. J. "The Original Spin Doctors: The Meeting of Perception and Insanity." *Perception* 34, no. 3 (2005): 253–260. doi:10.1068/p3403ed

Counter-Irritation

Gillies, H. Cameron. *The Theory and Practice of Counter-Irritation*. London: Macmillan, 1895.

Thomson, Spencer, and Henry H. Smith. *A Dictionary of Domestic Medicine and Household Surgery*. Philadelphia: Lippincott, Grambo, 1853.

Wand-Tetley, J. I. "Historical Methods of Counter-Irritation." *Rheumatology* 3, no. 3 (1956): 90–98. doi:10.1093/rheumotology/III.3.90

Surgery

Bishop, W. J. *The Early History of Surgery*. London: Hale, 1960.

Hollingham, Richard. *Blood and Guts: A History of Surgery*. New York: St. Martin's Press, 2009.

———. "With a Rusty Old Saw Like This, a Victorian Surgeon Could Amputate Your Leg in 30 Seconds Flat." *Mail Online*. August 22, 2008. www.dailymail.co.uk/femail/article-1045755/With-rusty-old-saw-like-Victorian-surgeon-amputate-leg-30-seconds-flat-One-snag--hadnt-invented-anaesthetic.html.

Leeches

Carter, Codell. "Leechcraft in Nineteenth-Century British Medicine." *Journal of the Royal Society of Medicine* 94 (January 2001): 38–42.

Fitzharris, Lindsey. "Lancets and Leeches and Cupping! Oh, My! Bloodletting Practices in Early Modern England." *The Chirurgeon's Apprentice.* November 23, 2011. http://thechirurgeons apprentice.com/2011/11/23/lancets-and-leeches-and-cup ping-oh-my-bloodletting-practices-in-early-modern-england/.

Tilt, Edward John. *A Handbook of Uterine Therapeutics and of Diseases of Women.* New York: Wood, 1868.

Hanging

Fitzharris, Lindsey. "News from the Dead: The Execution and Resuscitation of Anne Green." *The Chirurgeon's Apprentice.* September 28, 2010. http://thechirurgeonsapprentice.com/2010/09/ 28/news-from-the-dead-the-execution-resuscitation-of-anne -green/.

Moore, Wendy. *The Knife Man.* London: Bantam Press, 2005.

Hemophilia

Ingram, G. I. "The History of Haemophilia." *Journal of Clinical Pathology* 29, no. 6 (1976): 469–479. doi:10.1136/jcp.29.6.469

Kerr, C. B. "The Fortunes of Haemophiliacs in the Nineteenth Century." *Medical History* 7, no. 4 (October 1963): 359–370.

Public Anatomy

Guerrini, A. "Anatomists and Entrepreneurs in Early Eighteenth-Century London." *Journal of the History of Medicine and Allied Sciences* 59, no. 2 (2004): 219–239. doi:10.1093/jhmas/59 .2.219

Eels

Finger, Stanley, and Marco Piccolino. *The Shocking History of Electric Fishes: From Ancient Epochs to the Birth of Modern Neurophysiology.* New York: Oxford University Press, 2011.

"German Naturalists, Electric Eels and Horse Fishing." Beachcombing's Bizarre History Blog. December 8, 2010. www .strangehistory.net/2010/12/08/german-naturalists-and-electric -eels/.

Burke and Hare

"1829: William Burke, Eponymous Body-Snatcher." Executed Today.com. January 28, 2011. www.executedtoday.com/2011/01/28/1829-william-burke-hare-eponymous-body-snatcher/.

Stuttering

Bobrick, Benson. *Knotted Tongues: Stuttering in History and the Quest for a Cure.* New York: Simon and Schuster, 1995.

Plaenkers, Tomas. "Speaking in the Claustrum: The Psychodynamics of Stuttering." *International Journal of Psychoanalysis* 80, no. 2 (1999): 239–256. doi:10.1516/0020757991598693

Solomon, Meyer. "Remarks Upon Dr. Coriat's Paper 'Stammering as a Psychoneurosis'—A Criticism." *Journal of Abnormal Psychology* 10, no. 2 (1915): 120–137. doi:10.1037/h0072214

Lightning

Allison, G. A. "Therapeutic Effects of Lightning Upon Cancer." *The Lancet,* January 10 (1880).

Gould, George M. *Medical Curiosities: Adapted from Anomalies and Curiosities of Medicine.* Philadelphia: Saunders, 1992.

Presidents/Garfield

Hirschhorn, N., R. G. Feldman, and I. A. Greaves. "Abraham Lincoln's Blue Pills: Did Our 16th President Suffer from Mercury Poisoning?" *Perspectives in Biology and Medicine* 44, no. 3 (2001): 315–332.

Millard, Candice. *The Destiny of the Republic: A Tale of Madness, Medicine and the Murder of a President.* New York: Doubleday, 2011.

Vadakan, Vibul V. "The Asphyxiating and Exsanguinating Death of President George Washington." *Permanente Journal* 8, no. 2 (2004): 76–79.

Hoo Loo

Dormandy, Thomas. *The Worst of Evils: The Fight Against Pain.* New Haven, CT: Yale University Press, 2006.

"Hoo Loo, the Unfortunate Chinese." *Bell's Weekly Messenger* (London), April 17 (1831).

Spontaneous Human Combustion

Bondeson, Jan. *A Cabinet of Medical Curiosities.* Ithaca, NY: Cornell University Press, 1997.

Ignaz Semmelweis

Nuland, Sherwin B. *The Doctors' Plague: Germs, Childbed Fever, and the Strange Story of Ignác Semmelweis.* New York: Norton, 2003.

Youngson, R. M., and Ian Schott. *Medical Blunders.* New York: New York University Press, 1996.

Electricity

de la Peña, Carolyn Thomas. *The Body Electric: How Strange Machines Built the Modern American.* New York: New York University Press, 2003.

Monell, S. H., ed. *A System of Electrotherapeutics: Electricity in Diseases of the Eye, Ear, Nose, and Throat; Electricity in Genitourinary Diseases; Therapeutics of Static Electricity; Electricity in Dentistry.* Scranton, PA: International Textbook, 1902.

Morus, Iwan. "Marketing the Machine: The Construction of Electrotherapeutics as Viable Medicine in Early Victorian England." *Medical History* 36, no. 1 (1992): 34–52.

Hysteria/Manual Sexual Stimulation

Maines, Rachel. *The Technology of Orgasm: "Hysteria," the Vibrator, and Women's Sexual Satisfaction.* Baltimore: Johns Hopkins University Press, 1998.

Margolis, Jonathan. *O: The Intimate History of the Orgasm.* New York: Grove Press, 2004.

Lord Brunton/Sulpheretted Hydrogen

Brunton, Thomas L. "Disorders of Digestion: Their Consequences and Treatments." *British Medical Journal* 1 (1885): 57–61.

The Doctor's Riot

Bell, Whitfield J. "Doctor's Riot, New York." *New York Academy of Medicine Bulletin* (1976): 327–329.

Bloodletting

Clutterbuck, Henry. *Lectures on Blood-Letting: Delivered at the General Dispensary, Aldersgate Street.* Philadelphia: Haswell, Barrington, and Haswell, 1839.

Davis, Audrey B., and Toby A. Appel. *Bloodletting Instruments in the National Museum of History and Technology.* Washington, DC: Smithsonian Institution Press, 1979.

Kuriyama, Shigehisa. "Interpreting the History of Bloodletting." *Journal of the History of Medicine and Allied Sciences* 50, no. 1 (1995): 11–46. doi:10.1093/jhmas/50.1.11

"Medical Antiques: Collecting Surgical and Bloodletting Items." www.medicalantiques.com.

Porter, Dorothy, and Roy Porter. *Patient's Progress: Doctors and Doctoring in Eighteenth-Century England.* Stanford, CA: Stanford University Press, 1989.

Post, A. C. "Curative Effects of Bloodletting." *Bulletin of the New York Academy of Medicine* 44, no. 9 (September 1968): 1187–1189.

Stern, Heinrich. *Theory and Practice of Bloodletting.* New York: Rebman, 1915.

The Stethoscope

Bird, Golding. "Advantages Presented by the Employment of a Stethoscope with a Flexible Tube." *London Medical Gazette* 1 (1840): 440.

Bluth, Edward I. "James Hope and the Acceptance of Auscultation." *Journal of the History of Medicine and Allied Sciences* 25, no. 2 (1970): 202–210. doi:10.1093/jhmas/XXV.2.202

"The Medical Officer's Stethoscope." *The Lancet*, November (1879): 819.

Reiser, Stanley Joel. *Technological Medicine: The Changing World of Doctors and Patients.* New York: Cambridge University Press, 2009.

Masturbation and Spermatorrhea

Englehardt, Tristram. "The Disease of Masturbation: Values and the Concept of Disease." *Bulletin of the History of Medicine* 48 (Summer 1974): 234–248.

"Facts, and Important Information from Distinguished Physicians and Other Sources: Showing the Awful Effects of Masturbation on Young Men." Boston: Redding, 1843.

Hodgson, D. "Spermatomania—The English Response to Lallemand's Disease." *Journal of the Royal Society of Medicine* 98, no. 8 (2005): 375–379. doi:10.1258/jrsm.98.8.375

Milton, John Laws. *On Spermatorrhoea: Its Results and Complications.* London: Hardwicke, 1872.

Rosenman, Ellen Bayuk. "Body Doubles: The Spermatorrhea Panic." *Journal of the History of Sexuality* 12, no. 3 (2003): 365–399. doi:10.1353/sex.2004.0013

Pain

Dormandy, Thomas. *The Worst of Evils: The Fight Against Pain.* New Haven, CT: Yale University Press, 2006.

Rabies

Carter, Codell. "Nineteenth-Century Treatments for Rabies as Reported in the *Lancet*." *Medical History* 26, no. 1 (January 1982): 67–78.

American Institutions for Idiotic and Feeble-Minded Persons

American Association on Mental Deficiency, comp. *Proceedings of the Association of Medical Officers of American Institutions for Idiotic and Feeble-Minded Persons.* Vols. 1–2. N.p.: General Books, 2010.

Medical Racism

Cartwright, Samuel. "Diseases and Peculiarities of the Negro Race." *DeBow's Review—Southern and Western States* 11 (1851).

Hossain, Shah A. "'Scientific Racism' in Enlightened Europe: Linnaeus, Darwin and Galton." January 16, 2008. http://serendip.brynmawr.edu/exchange/node/1852.

Jarvis, E. "Insanity Among the Colored People of the United States." *American Journal of Insanity*, January (1852): 268–282.

Teething

Farnsworth, Diane. "The Causes of Pink Disease." Pink Disease Support Group. www.pinkdisease.org/causePD.htm.

"Treatments for Children: Teething." Royal Pharmaceutical Society. www.rpharms.com/museum-pdfs/g-teethingtreatments.pdf.

Bibliopegy

Fitzharris, Lindsey. "Books of Human Flesh: The History Behind Anthropodermic Bibliopegy." *The Chirurgeon's Apprentice*. January 31, 2012. http://thechirurgeonsapprentice.com/2012/01/31/books-of-human-flesh-the-history-behind-anthropodermic-bibliopegy/.

Jacobs, Samuel. "The Skinny on Harvard's Rare Book Collection." *The Harvard Crimson*, February 2 (2006).

Vivisection

Fitzharris, Lindsey. "Dissecting the Living: Vivisection in Early Modern England." *The Chirurgeon's Apprentice*. August 29, 2011. http://thechirurgeonsapprentice.com/2011/08/29/dissecting-the-living-vivisection-in-early-modern-england/.

Gillispie, Charles Coulston. *Complete Dictionary of Scientific Biography*. Detroit, MI: Scribner, 2008.

"A History of Antivivisection from the 1800s to the Present: Part I (mid-1800s to 1914)." *The Black Ewe*. http://brebisnoire.wordpress.com/a-history-of-antivivisection-from-the-1800s-to-the-present-part-i-mid-1800s-to-1914/.

Doctor/Patient Modesty

Brumberg, Joan Jacobs. *The Body Project: An Intimate History of American Girls*. New York: Random House, 1997.

Nicolson, Malcolm. "The Art of Diagnosis: Medicine and the Five

Senses." In W. F. Bynum and Roy Porter, eds., *Companion Encyclopedia of the History of Medicine*, Vol. 2 (London: Routledge, 1993), 801–825.

Cautery (Modern)

Seguin, Edward. "Use of the Actual Cautery in Medicine." *Archives of Medicine* 1 (April 1879): 312.

Bed-Wetting

Salman, Michael A. "An Historical Account of Nocturnal Enuresis and Its Treatment." *Proceedings of the Royal Society of Medicine* 68, no. 7 (July 1975): 443–445.

Schultheiss, Dirk. "A Brief History of Urinary Incontinence and Its Treatment." *European Urology* 38, no. 3 (September 2000): 352–362.

John Hunter

Moore, Wendy. *The Knife Man*. London: Bantam Press, 2005.

Buried Alive

Allen, George W. *Premature Burial: An Argument Before a Legislative Committee on a Bill "to Provide for the Resuscitation of Those Apparently Dead and to Prevent Premature Encoffinment, Burial, or Cremation."* Boston: Ellis, 1905.

"The Fear of Premature Burial." *The Lancet* 144, no. 3701 (1894): 265–266. doi:10.1016/S0140-6736(01)58627-4

Fitzharris, Lindsey. "Torturing the Dead: The Prevention of Premature Burial and Dissection." *The Chirurgeon's Apprentice*. April 4, 2010. http://thechirurgeonsapprentice.com/2012/04/10/torturing-the-dead-the-prevention-of-premature-burial-and-dissection/.

Tebb, William, and Edward Perry Vollum. *Premature Burial, and How It May Be Prevented, with Special Reference to Trance Catalepsy, and Other Forms of Suspended Animation*. London: Swan, 1905.

Thompson, H. "Premature Burial." *The Lancet* 149, no. 3844 (1897): 1235–1236. doi:10.1016/S0140-6736(01)95910-0

Walsh, David. *Premature Burial: Fact or Fiction?* New York: Wood, 1898.

Tapeworm Trap

Caro. "Something to Show and Scare the People." *The Quack Doctor.* August 25, 2010. http://thequackdoctor.com/index.php/something-to-show-and-scare-the-people/.

John Oleson/Patent Medicines

Oleson, Charles Wilmot. *Secret Nostrums and Systems of Medicine: A Book of Formulas.* Chicago: Oleson, 1891.

The Tongue

Haller, John S. "The Foul Tongue: A 19th-Century Index of Disease." *Western Journal of Medicine* 137 (1982): 258–260.

Ridge, Benjamin. *Glossology: Or the Additional Means of Diagnosis of Disease to Be Derived from Indications and Appearances of the Tongue: Read Before the Senior Physical Society of Guy's Hospital, 4th November, 1843.* London: Churchill, 1844.

Dr. Arbuthnot Lane—Colon Surgery

Dally, Ann G. *Fantasy Surgery: 1880–1930: With Special Reference to Sir William Arbuthnot Lane.* Amsterdam: Rodopi, 1996.

Sullivan-Fowler, Micaela. "Doubtful Theories, Drastic Therapies: Autointoxication and Faddism in the Late Nineteenth and Early Twentieth Centuries." *Journal of the History of Medicine and Allied Sciences* 50, no. 3 (1995): 364–390.

Shoe Fluoroscopy

A Guide for Uniform Industrial Hygiene Codes or Regulations for the Use of Fluoroscope Shoe Fitting Devices. Technical paper. American Conference of Governmental Industrial Hygienists, 1951.

Nedd, Council A. "When the Solution Was the Problem: A Brief History of the Shoe Fluoroscope." *American Journal of Roentgenology* 158, no. 6 (1992): 1270.

Dr. Cotton, Electroshock Therapy, Insulin Shock Therapy, Lobotomy

Goodman, Barak, and John Maggio, dirs. *The Lobotomist*. Performed by Walter Freeman. *American Experience*. www.pbs.org/wgbh/americanexperience/films/lobotomist/.

Porter, Roy. *Madness: A Brief History*. Oxford, UK: Oxford University Press, 2002.

Scull, Andrew T. *Madhouse: A Tragic Tale of Megalomania and Modern Medicine*. New Haven, CT: Yale University Press, 2005.

Valenstein, Elliot S. *Great and Desperate Cures: The Rise and Decline of Psychosurgery and Other Radical Treatments for Mental Illness*. New York: Basic Books, 1986.

Whitaker, Robert. *Mad in America: Bad Science, Bad Medicine, and the Enduring Mistreatment of the Mentally Ill*. Cambridge, MA: Perseus, 2002.

Floating Kidneys

Culbertson, William. "Displaced and Movable Kidney in Women: Its Symptomatology, Diagnosis and Treatment." *Canadian Medical Association Journal*, August (1912): 985–994.

Dally, Ann G. *Fantasy Surgery: 1880–1930: With Special Reference to Sir William Arbuthnot Lane*. Amsterdam: Rodopi, 1996.

Suckling, Cornelius William. *Movable Kidney: A Cause of Insanity, Headache, Neurasthenia, Insomnia, Mental Failure and Other Disorders of the Nervous System. A Cause Also of Dilatation of the Stomach*. London: Lewis, 1905.

Penicillin

Diggins, F. W. "The True History of the Discovery of Penicillin, with Refutation of the Misinformation in the Literature." *British Journal of Biomedical Science* 56, no. 2 (1999): 83–93.

Goldsworthy, P. D., and A. C. McFarlane. "Howard Florey, Alexander Fleming and the Fairy Tale of Penicillin." *Medical Journal of Australia* 176, no. 4 (2002): 176–178.

Lehrer, Steven. *Explorers of the Body*. Garden City, NY: Doubleday, 1979.

Index

masturbation causing, 132
medieval medicine treatment of, 20–21
nostalgia as, 66
Renaissance medicine treatment of, 65
Rush treatment contributions to, 95–97
shock therapy for, 173
slavery and, 140–41
surgery suffered by those with, 167, 169–70, 176–78
swinging in treatment of, 96–97, 150
transorbital lobotomy in, 177–78
wind's part in, 150
Mercury, 68–69, 116–17
Metchnikoff, Elie, 166, 167
Microscope and microscopic work, 80–82
Midwives, 76–77. *See also* Man-midwives
Milton, John Laws, 145
Moliére, 46
Monell, S. H., 123
Money, 37–39
Moniz, Egas, 177–78
Monstrorium Historia (Aldronvandi), 71
Morton, W. T. G., 98
Mortuaries, 157
Murder Act, 106
Music, 85–87
Mutter Museum of the College of Physicians of Philadelphia, 72–73

Nausea, 64–65
Nero, 13, 69
New England Journal of Medicine, 170
New York Journal of Medicine, 115
New York Times, 170
Nightingale, Florence, 102
Nihill, Elizabeth, 77

Nose, 61–63
Nostalgia, 65–66

Observations Medicorum Rariorum (Schenkius), 70
Oleson, Charles, 160–61
On Hemorrhoids (Hippocrates), 6
On Urine (de Corbeil), 27
Oral fixation, 113–14, 114n4
Organ removal, 167, 169–70

Pain
    African Americans and, 140
    chirugeons and necessary, 18
    medical tradition of, 135–36
    patient and surgical, 135–36
    sin causing, 136
Palpation (touching), 129, 148–49
Paracelsus, 51, 52–53, 58
Paré, Ambroise
    bosum serpent approach of, 69
    demonic possession self-diagnosis by, 84
    on dental work, 73
    maternal impressions for, 55
    poison study on living person by, 89–90
    surgery revolutionized by, 45
Pasteur, Louis, 158
Patients
    bosum serpents in bellies of, 69–70
    heroic doctor palpation of, 129, 148–49
    insulin comas for, 175
    medieval physician bedside manner for, 28–30
    medieval surgeons and avoiding desperate, 38
    money and medieval surgeon payment from, 37–39
    pain of, 135–36
    spinning of, 99–100, 150
    surgical procedures and restraint of, 19
Penicillin, 174–75

Uline Ice Company, 177
University of Paris, 16
"Untoward Result of Drunkenness, An" (Provincial Medical and Surgical Journal), 72
Urine, 26–28, 41
U.S. presidents, 116–18

van Helmont, Jan Baptist, 91
Vesalius, 45, 56–57, 143
Viennese Medical Journal, 121
Violence, 38
von Humboldt, Alexander, 110–11
von Hutten, Ulrich, 67, 68
von Pfolspeundt, Heinrich, 62

Warm Beer: Or a Treatise Wherein It Is Declared That Beer So Qualified Is Far More Wholesome Than That Which Is Drank Cold, 82, 83
Warts, 11
Washington, George, 116
Weapon salve, 49–51
Wells, Horace, 98
Whitney, Henry Clay, 117
William of Saliceto, 19, 21–22, 34

Wind, 150
Witchcraft, 83–85
Womb, 9, 39
Women
    birth and birthing as work of, 76
    "floating kidneys" in upper-class, 171
    Greek doctors on womb of, 9
    heroic doctor palpation of, 148–49
    leeches risk to, 105
    medieval anatomical and reproductive texts for, 39–40
    menstruation, 10, 36, 40
Wootton, David, ix
Worm, Ole, 71, 91
Wounds, 25–26, 34–35

X-ray and radiation, 167–69

Yperman, Jehan, 34–35, 40–41

Al-Zaharawi, Abu al-Qasim Khalaf ibn al-Abbas, 22
Zeigler of Strasbourg, 90–91
Zerbi, Gabriel, 30
Zoonomia (Darwin), 99